4th Symposium on the Mechanics of Slender Structure

(MoSS2013)

AA002703

Journal of Physics: Conference Series Volume 448

Harbin, China
7 – 9 January 2013

ISBN: 978-1-62993-073-2
ISSN: 1742-6588

Printed from e-media with permission by:

Curran Associates, Inc.
57 Morehouse Lane
Red Hook, NY 12571

Some format issues inherent in the e-media version may also appear in this print version.

Copyright© by the Institute of Physics
All rights reserved.

Printed by Curran Associates, Inc. (2013)

For permission requests, please contact the Institute of Physics
at the address below.

Institute of Physics
Dirac House, Temple Back
Bristol BS1 6BE UK

Phone: 44 1 17 929 7481
Fax: 44 1 17 920 0979

techtracking@iop.org

Additional copies of this publication are available from:

Curran Associates, Inc.
57 Morehouse Lane
Red Hook, NY 12571 USA
Phone: 845-758-0400
Fax: 845-758-2634
Email: curran@proceedings.com
Web: www.proceedings.com

TABLE OF CONTENTS

Journal of Physics: Conference Series **448** (2013) 011001 doi:10.1088/1742-6596/448/1/011001

Preface

This volume of *Journal of Physics: Conference Series* contains papers presented at the 4th Symposium on the Mechanics of Slender Structures (MoSS2013) run under the auspices of the Institute of Physics Applied Mechanics Group and hosted by Harbin Institute of Technology (China) from 7–9 January 2013. The conference has been organized in collaboration with the Technical Committee on Vibration and Sound of the American Society of Mechanical Engineers and follows a one day seminar on *Ropes, Cables, Belts and Chains: Theory and Applications* and the MoSS2006 symposium held at the University of Northampton (UK) in 2004 and 2006, respectively, the MoSS2008 symposium held at the University of Maryland Baltimore County (USA) in 2008 and the MoSS2010 symposium hosted by Mondragon University and held in San Sebastian (Spain) in 2010.

The remit of the Symposium on the Mechanics of Slender Structures series involves a broad range of scientific areas. Applications of slender structures include terrestrial, marine and space systems. Moving elastic elements such as ropes, cables, belts and tethers are pivotal components of many engineering systems. Their lengths often vary when the system is in operation. The applications include vertical transportation installations and, more recently, space tether propulsion systems. Traction drive elevator installations employ ropes and belts of variable length as a means of suspension, and also for the compensation of tensile forces over the traction sheave. In cranes and mine hoists, cables and ropes are subject to length variation in order to carry payloads. Tethers experiencing extension and retraction are important components of offshore and marine installations, as well as being proposed for a variety of different space vehicle propulsion systems based on different applications of momentum exchange and electrodynamic interactions with planetary magnetic fields. Furthermore, cables and slender rods are used extensively in civil engineering; in cable-supported bridges, guyed masts and long-span roofs of buildings and stadia. Suspended cables are also applied as electricity transmission lines. Chains are used in various power transmission systems that include such mechanical systems as chain drives and chain saws. Moving conveyor belts are essential components in various material handling installations. Other structures such as pipelines, plates, beams, mechanical linkages and DNA structures also fall into this category.

The behaviour of these elements plays a significant role in the performance of the host structure and a holistic approach is needed in the analysis and design of the entire system. This meeting brings together experts from various fields: structural mechanics, thermo-mechanics, dynamics, electrodynamics, vibration and control, structural health monitoring, artificial intelligence, materials science and applied mathematics to discuss the current state of research as well as rising trends and direction for future research in the area of mechanics of slender structures. The event is aimed at improving the understanding of structural and thermo-mechanical properties and behaviour of slender structures. The papers presented at the conference cover analytical, numerical and experimental research in various applications of slender structures.

The Organizing Committee gratefully acknowledges support received from the co-sponsoring institutions and would like to thank the authors for their hard work and high quality contributions.

Dengqing Cao
Harbin Institute of Technology

Stefan Kaczmarczyk
The University of Northampton

Some Aspects of Research on Mechanics of Thin Elastic Rod

Liu Yanzhu

Department of Engineering Mechanics, Shanghai Jiao Tong University, Shanghai 200240, China

liuyzhc@gmail.com

Xue Yun

School of Mechanical Engineering, Shanghai Institute of Technology, Shanghai 201418, China

xy@sit.edu.cn

Abstract. Some aspects of research wok on the mechanics of thin elastic rod based on Kirchhoff-Cosserat's model were summarized. The analytical mechanics with arc-coordinate s and time t as double variables was established to formulate the motion of elastic rod. In stability analysis the difference and relationship between Lyapunov's stability and Euler's stability were discussed. The first approximate stability was determined by the characteristic equation with double eigenvalues in different domains, one of which can be determined by geometric conditions in static analysis. The Lyapunov's and Euler's stability conditions of the rod in space domain are the necessary conditions of Lyapunov's stability in time domain. As applications of the Kirchhoff's rod in molecular biology, the explanations of nucleosome structure and the chromosome coiling of DNA were given. Concerning the application in engineering the shape of a hanging rod under gravity and the coiling and stretching process of an extendable space mast were discussed. The motion of an axial moving beam with constant velocity and axial extensive force was discussed as an example of exact Cosserat's rod, the special case of small deformation is the Timoshenko's beam.

1. Introduction

The mechanics of a thin-long elastic rod has a practical background in molecular biology and engineering. The traditional model of a beam using Cartesian coordinates of centerline cannot be competent in analysis of a thin-long rod with large deformation. The Kirchhoff's kinetic analogy provides a different model of the rod by rotation of cross section along the centerline, and the attitude angles of cross section are used as unknown functions of arc-coordinates s. In analysis of dynamics the time variable is added and the model becomes a system with double independent variables s and t. The Cosserat's model is an exacter model, which abandons the hypothesis of inextensible and unshearable characters, and is closer to a real rod in practice. The Kirchhoff-Cosserat's model is suitable particularly to the large deformation of thin-long elastic rod, such as DNA, fiber, sub-ocean cable, oil-well drill string and other thin-long projects. In application of Kirchhoff-Cosserat's model all approaches of dynamics, such as the dynamics of rigid bodies, the analytical mechanics, the theory of stability and others, can be used into the mechanics of elastic rod. In this paper some aspects of research on Kirchhoff-Cosserat's model are summarized.

2. Analytical mechanics of elastic rod [1~6]

As basic definitions in analytical mechanics of elastic rod, the virtual displacement of cross section of the rod is defined as: The imagined infinitesimal displacement of cross section, consistent with constraint, and irrelevant to the arc-coordinates and time. The principle of D'Alembert-Lagrange is defined as: For an elastic rod under ideal bilateral constraint, the true motion different to any possible motion consistent with constraint, that its virtual work for arbitrary virtual displacement is zero. The principle of minimal potential energy is used to take place of Hamilton principle, and the density of potential energy Γ of deformation of the rod plays a role of Lagrange's function in analytical statics.

$$\delta\int_0^L \Gamma \,\mathrm{d}s = 0, \quad \Gamma = \frac{1}{2}\left[A\omega_1^2 + B\omega_2^2 + C\left(\omega_3 - \omega_3^0\right)^2\right] - F\gamma \tag{2.1}$$

When the rod under constraint condition $\boldsymbol{C}(\boldsymbol{q},\dot{\boldsymbol{q}}) = 0$, the Lagrange's multiplier $\boldsymbol{\Lambda}$ can be used to compose a revised Lagrange's function $\tilde{\Gamma} = \Gamma + \boldsymbol{\Lambda}\cdot\boldsymbol{C}$. The Lagrange's equations are derived as

$$\frac{\mathrm{d}}{\mathrm{d}s}\left(\frac{\partial\tilde{\Gamma}}{\partial q_i'}\right) - \frac{\partial\tilde{\Gamma}}{\partial q_i} = 0 \quad (j = 1, 2, \cdots, n) \tag{2.2}$$

In dynamic analysis it is necessary to establish an analytical dynamics with double arguments s and t. Using the Lagrange's function $\Lambda = \Gamma - T$, composed of densities of potential energy Γ and kinetic energy T, the Lagrange's equations and the Nielson's equations can be derived as

$$\frac{\partial}{\partial s}\left(\frac{\partial\Lambda}{\partial q_j'}\right) + \frac{\partial}{\partial t}\left(\frac{\partial\Lambda}{\partial\dot{q}_j}\right) - \frac{\partial\Lambda}{\partial q_j} + f_j = 0 \quad (j = 1, 2, \cdots, n) \tag{2.3}$$

$$\frac{\partial\Lambda'}{\partial q_j'} + \frac{\partial\dot{\Lambda}}{\partial\dot{q}_j} - 3\frac{\partial\Lambda}{\partial q_j} + f_j = 0 \quad (j = 1, 2, \cdots, n) \tag{2.4}$$

3. Stability and vibration [7~17]

When the Lyapunov's stability theory is applied to analyze the equilibrium of elastic rod some unusual results can be obtained. As an example the equilibrium of an axial compressed elastic rod is stable by Lyapunov's stability definition (Fig.1a), and is unstable when the rod is extended (Fig.1b). The results contradict to the traditional knowledge of Euler's buckling concept of a compressed bar evidently. The different conclusions are caused by different definitions of stability. According to the Lyapunov's stability theory, the equilibrium is stable when the perturbed centerline is restricted in a small neighborhood of the unperturbed straight line. Nevertheless, from the viewpoint of Euler's stability, when the perturbation equations have nontrivial solution satisfying the boundary conditions on both ends, the equilibrium is unstable and the buckling occurs. The Lyapunov's stability is a concept of stability with respect to the initial perturbation, but the perturbed centerline is under constraints on both ends. When all boundary conditions on both ends are satisfied, the corresponding load is the Euler's critical load. Therefore the Euler's load can be calculated in analysis of Lyapunov's stability.

(a) Compressed rod (stable) (b) Extended rod (unstable)

Figure 1. Perturbed centerlines of compressed and extended rod

In stability analysis the linearized perturbed equations has double arguments s and t, it follows that the characteristic equation has double eigenvalues λ and w corresponding to different variables. When the Lyapunov's stability conditions in statics are satisfied, the eigenvalue in space domain $\lambda = \pm ik$ can be determined by boundary conditions on both ends. Then the characteristic equation contains only one unknown eigenvalue w, which determines the stability of equilibrium in time domain.

As an example we consider an elastic rod with circular cross section and length L. The dynamical equations of a Kirchhoff's rod with the Euler's angles ψ, ϑ, φ and internal forces F_i $(i=1,2,3)$ as unknown variables permit special solutions, corresponding to the helical equilibrium

$$\vartheta = \vartheta_0, \psi = \omega_0 s, \phi = 0, F_1 = 0, F_2 = F_0 \sin\vartheta_0, F_3 = F_0 \cos\vartheta_0 \tag{3.1}$$

where F_0 is the axial force required by the helical equilibrium

$$F_0 = -\frac{EI\omega_0^2}{\cos\vartheta_0} \tag{3.2}$$

where E, I are Young's modulus and inertial moment of the cross section. The characteristic equation of the linearized perturbed equations with perturbations x_i $(i=1,2,\cdots,6)$ can be derived as

$$a(\lambda)w^4 + b(\lambda)w^2 + c(\lambda) = 0 \tag{3.3}$$

where $c(\lambda)$ is defined as

$$c(\lambda) = \lambda^4\left(1+\lambda^2\right)^2\left[1+\left(\lambda^2+3\right)\cos^2\vartheta_0\right] \tag{3.4}$$

The characteristic equation in space domain $c(\lambda) = 0$ has pure imaginary roots $\lambda = \pm ik$, where

$$k = \left(\frac{1+3\cos^2\vartheta_0}{\cos^2\vartheta_0}\right)^{1/2} \tag{3.5}$$

Therefore the helical equilibrium is stable by Lyapunov's definition. The boundary conditions $x_i(L) = x_i(0)$ on both ends of the rod require $k = 2n\pi/\omega_0 L$ $(n=1,2,\cdots)$, then the Euler's critical load can be derived from Eqs.(3.5) and (3.2) as

$$\left|F_0\right|_{\mathrm{cr}} = \frac{4n^2\pi^2 EI\cos\vartheta_0}{L^2\left(1+3\cos^2\vartheta_0\right)} \tag{3.6}$$

After substitution of the determined eigenvalue $\lambda = \pm ik$ into Eq.(3.3), one of the pure root conditions of eigenvalue w is $c(\lambda) > 0$. Comparing it to the characteristic equation $c(\lambda) = 0$, it follows that the axial force should be smaller than the Euler's critical load. Thus the Lyapunov's and Euler's stability conditions of the helical rod in space domain are the necessary conditions of Lyapunov's stability in time domain. When all stability conditions are satisfied letting $w = \pm i\mu$, we obtain the angular frequency μ of lateral vibration of the helical rod as

$$\mu = \left(\frac{2n\pi}{L}\right)^2\sqrt{\frac{EI}{\rho S}}\,\Phi(k,\vartheta_0) \tag{3.7}$$

$$\Phi(k,\vartheta_0) = \sqrt{\frac{k^2+1+4\cos^2\vartheta_0}{k^2+\sin^2\vartheta_0}}\left\{1\mp\sqrt{1+\frac{\left(k^2+\sin^2\vartheta_0\right)\left(k^2-1\right)^2\left[1-\left(k^2-3\right)\cos^2\vartheta_0\right]}{k^4\left(k^2+1+4\cos^2\vartheta_0\right)^2\cos^2\vartheta_0}}\right\}^{1/2} \tag{3.8}$$

4. Application in molecular biology [18~20]

4.1 Explanation of the nucleosome structure of DNA

As the structural unit of chromatin the nucleosome is composed of DNA and histone core HO with a small H1-histone (Fig.2, Cited from http://www.cvh.cc/onews.asp?id=29713). The DNA wraps around histones 1 and 3/4 circle and forms a nucleosome structure. Many nucleosomes are linked together like a beads. The nucleosome structure can be simplified as a thin elastic rod constrained by a cylinder with radius R and locked by a small bar at the entrance of DNA (Fig.3).

In order to explain the structure of nucleosome, at first we consider the equilibrium of the thin rod acted by the electric force f_e and contact force f_c with constraint conditions

$$\vartheta = \vartheta_0, \ \psi' = \frac{\sin\vartheta_0}{R}, \ \varphi' = \omega_0 \tag{4.1}$$

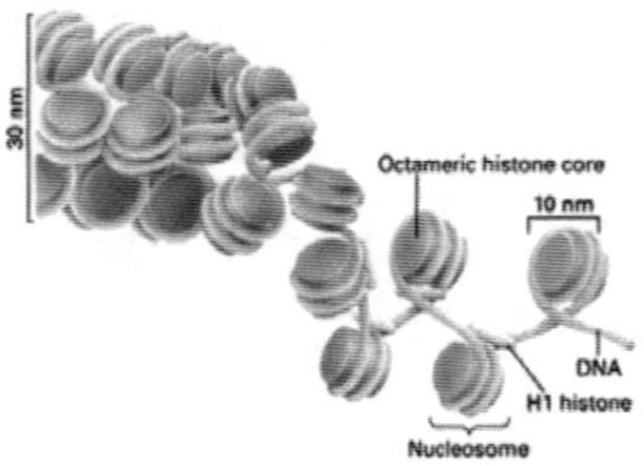

Figure 2. Nucleosome structure

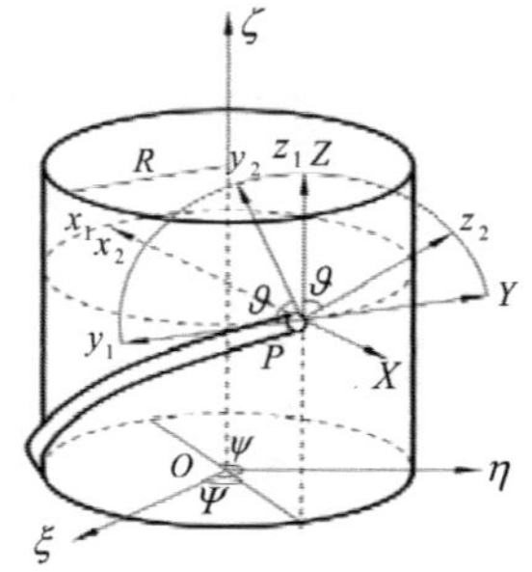

Figure 3. Thin rod constrained by cylinder

The internal forces and the contact forces can be solved as

$$\begin{gathered} F_1 = 0, F_2 = -\frac{A-C}{R^2}\sin^3\vartheta_0\cos\vartheta_0, F_3 = F_{30} \\ f_{c1} = \left(F_{30} + \frac{A-C}{4R^2}\sin^2 2\vartheta_s\right)\frac{\sin^2\vartheta_s}{R} + f_e, \ f_{c2} = f_{c3} = 0 \end{gathered} \tag{4.2}$$

where $A = EI$ and $C = GI_0$ are the bending and twisting stiffness of the rod. Considering the equilibrium of the cylinder HO wrapped by the rod we obtain the contact range of DNA

$$[\psi_1, 4\pi - \psi_1] = [102.5^\circ, 617.5^\circ] \tag{4.3}$$

After numerical calculation the range is about 1.43 circles (Fig.4). At last we consider the equilibrium of the H1-histone, which is simplified as a small bar in contact with the cylinder HO and DNA. The incline angle β of the bar can be obtained as

$$\tan\beta = -\frac{\mu R^3\cos\psi_1\tan\vartheta_0}{\mu R^3 + (C-A)\sin^4\vartheta_0} \tag{4.4}$$

where $\mu = f_{c1} - f_e$. When the electric force is balanced by the contact force, $\mu = 0$, and $\beta = 0$, the H1 bar is erect along ζ-axis (Fig.5).

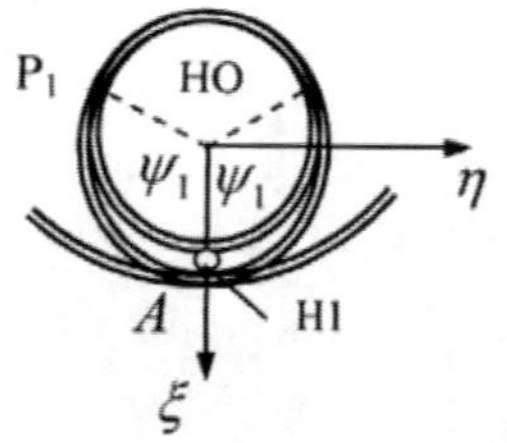

Figure 4. Contact range of DNA

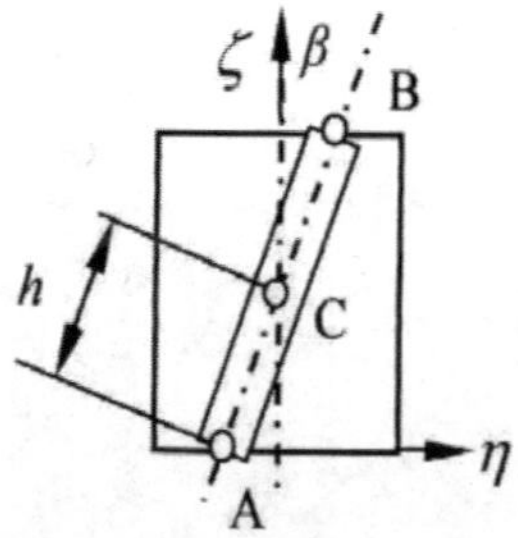

Figure 5. Equilibrium of H1-bar

4.2 Explanation of chromosome coiling of DNA

The formation of DNA in chromatin is coiled in helical state with different order. At first the nucleosome structure folds the DNA to a helix with diameter 11 nm, after that the helix twins itself in higher order coiling with diameter 30 nm into the chromatin (Fig.6, Cited from http://www. cchem. berkeley.edu/jehgrp/yms/). In order to explain the coiling process, a physical experiment was made by Thompson and Champney using a rubber rod under axial tension force F_0 and twist M_0. When the force F_0 increases the initial straight equilibrium becomes unstable and transforms to a helical equilibrium. When F_0 increases further the straight helical rod becomes unstable and twins to a helix with higher order. In explanation of the phenomenon we write the equilibrium equations of a Kirchhoff's rod under axial tension and twist, and uncouple it to a nonlinear equation of ϑ

$$\frac{d^2\vartheta}{ds^2}+Q(\vartheta)=0,\quad Q(\vartheta)=\frac{l^2\sin\vartheta}{4}\left[\left(\cos\frac{\vartheta}{2}\right)^{-4}-\frac{2p}{l^2}\right] \tag{4.5}$$

Where $p=2F_0/A$ and $l=M_0/A$. The trivial solution $\vartheta_0=0$ or π of equation $Q(\vartheta_0)=0$ corresponds to a straight rod. In addition, when $2p/l^2>1$ a nontrivial solution exists as

$$\vartheta_0=2\arccos\left[\left(2p/l^2\right)^{-1/4}\right] \tag{4.6}$$

The solution (4.6) describes a helical equilibrium with pitch angle $\pi/2-\vartheta_0$, which decreases with increase of $2p/l^2$. The stability analysis confirms that ϑ_0 has a bifurcation point of $2p/l^2=1$, after which the stable straight state becomes unstable and a stable helical state exists (Fig.7). When F_0 increases further the straight thin helical rod becomes unstable and twins to a coiling state of 2-nd order. The bifurcation point occurs on $2\tilde{p}/\tilde{l}^2=1$, where $\tilde{p}=2F_0/\tilde{A}$, $\tilde{l}=M_0/\tilde{A}$, and $\tilde{A}$ is an equivalent bending stiffness of a thin helical rod

$$\tilde{A}=A\cos\vartheta_0\left[1+\frac{(1-\lambda)}{2\lambda}\sin^2\vartheta_0\right]^{-1},\quad \lambda=\frac{C}{A} \tag{4.7}$$

Continuing the similar process the forming of different order helix in chromosome coiling can be explained qualitatively.

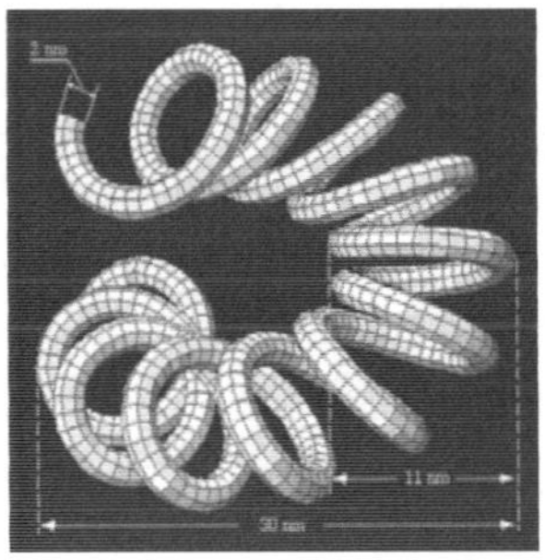

Figure 6. Chromosome coiling of DNA

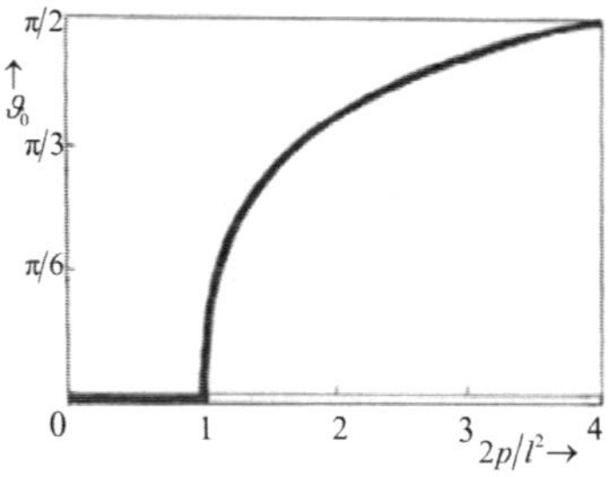

Figure 7. Bifurcation of ϑ_0 vs $2p/l^2$

5. Application in engineering [21~24]

5.1 Shape of hanging rod under gravity

The shape of a flexible cord under gravity was expressed by hyperbolic function in seventeen century as a successful application of mathematical calculus. The well-known curve of catenary can be utilized in engineering as an approximate model of hanging cable or pipe. When the hanging cable has large cross section the influence or bending rigidity cannot be neglected. The equilibrium equation of Kirchhoff's rod under gravity can be written with consideration of the bending rigidity

$$EI\vartheta'' + (\rho Sgs)\cos\vartheta - F_0\sin\vartheta = 0 \tag{5.1}$$

Where ϑ is the incline angle of the tangent of centerline to the horizon, ρ and S are the density and area of cross section (Fig.8). Using the perturbation method with a small parameter $\varepsilon = EI/mgl^2$ and taking the traditional solution of catenary as zero-th approximate solution, where l is the length of the rod, we obtain the solution corrected by the bending stiffness

$$\vartheta(s) = \arctan(\beta s) - \frac{2\varepsilon\beta^4 l^3 s}{(1+\beta^2 s^2)^{3/2}} \tag{5.2}$$

where $\beta = \rho Sg/F_0$. Introduce a function $\phi(\beta x)$

$$\phi(\beta x) = 2\beta^3 l^3 \int_0^{\beta x} \frac{du}{\mathrm{ch}^3 u} \tag{5.3}$$

The shape of the hanging rod can be derived in analytical form as

$$y(x) = \frac{1}{\beta}\left[\mathrm{ch}(\beta x) - 1 - \varepsilon\int_0^{\beta x} \mathrm{ch}u\,\phi(u)\,du\right] \tag{5.4}$$

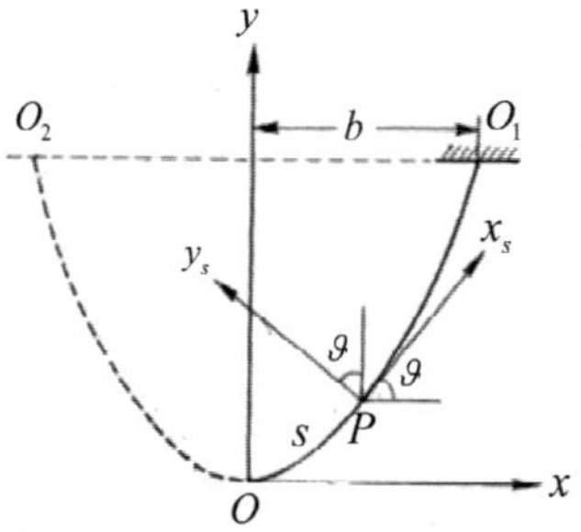

Figure 8. The hanging rod under gravity

5.2 Coiling and stretching of extendable space mast

The coiling and stretching process of the main element of extendable space mast in astronautic technique can be simplified by the motion of an elastic rod under unilateral constraint of a cylinder (Fig.9). The large deformation from a straight rod to a flat helical rod can be analyzed by the Kirchhoff's model with consideration of distributed contact force. In order to simplify the problem we assume that the rod maintains helical shape with unchanged radius R as the result of cylindrical constraint, and only the variation of angle ϑ and twisting of the rod are considered. In discussion of coiling process the inertial effect can be neglected approximately when the deformation is slow enough. The necessary external forces and torques in coiling process can be determined by some first integrals as

$$F(\vartheta) = \frac{A(1-\lambda)}{R^2}\cos\vartheta\sin^2\vartheta,\quad M(\vartheta) = \frac{A\sin\vartheta}{R}(\lambda\cos^2\vartheta + \sin^2\vartheta) \tag{5.5}$$

The stretching of the rod in space is a fast process, and the inertial effect should be considered. Based on the hypothesis of maintaining helical shape, the dynamical equations can be simplified to a nonlinear equation of $\vartheta(t)$

$$\ddot{\vartheta} - g(\vartheta)\dot{\vartheta}^2 = 0,\quad g(\vartheta) = \varepsilon\cos\vartheta\sin\vartheta(1+\cos^2\vartheta) \tag{5.6}$$

where $\varepsilon = (a/2R)^2$, a is the diameter of the rod. Eq.(5.6) permits a first integral to determine the phase trajectories in phase plane $(\vartheta,\dot{\vartheta})$ (Fig.9)

$$\dot{\vartheta}(\vartheta)=\dot{\vartheta}_0\left\{1+\frac{\varepsilon}{4}\left[2\left(\cos^2\vartheta_0-\cos^2\vartheta\right)+\cos^4\vartheta_0-\cos^4\vartheta\right]\right\} \tag{5.7}$$

Thus the stretching velocity $v_\zeta(s)$ and the time t_{str} of the duration from a flat helix to a straight rod are obtained in analytical form

$$v_\zeta=-\left(\sqrt{\frac{E}{\rho}}\dot{\vartheta}\sin\vartheta\right)s,\quad t_{\mathrm{str}}=\frac{\pi S l^2}{4I}\sqrt{\rho E}\left(1+\frac{11\varepsilon}{32}\right) \tag{5.8}$$

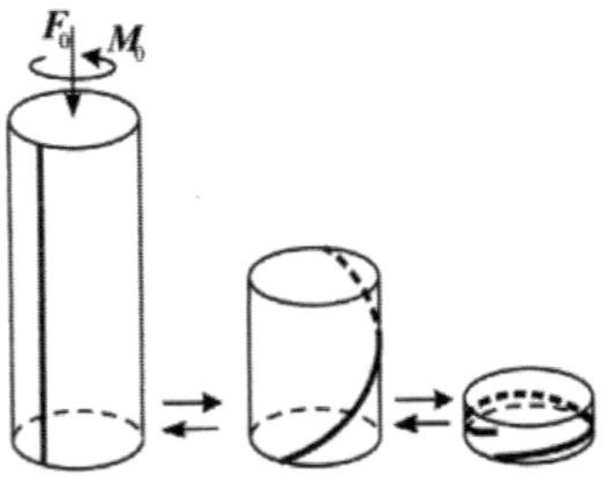

Figure 9. Coiling and stretching of a rod

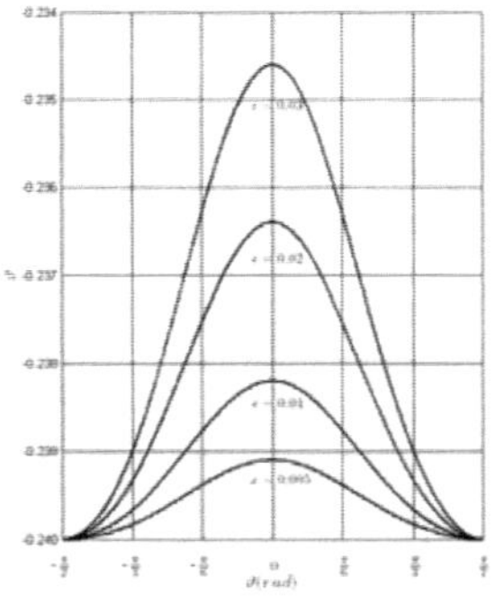

Figure 9. Phase trajectories in $(\vartheta,\dot{\vartheta})$ plane

6. Exact Cosserat's model [25~28]

The inextensible and unshearable hypotheses of Kirchhoff's model are not adaptable to the real soft materials. The exact Cosserat's model takes the extension of centerline and the shear deformation of cross section in consideration, and is closer to a real elastic rod in practice. As the result of axial extension the arc-coordinate s of an arbitrary point P in centerline is changed by $s^*=s(1+\varepsilon_3)$, where ε_3 is the normal strain of centerline. The shear strains $\varepsilon_1,\varepsilon_2$ of cross section causes an additional rotation of the tangent vector of centerline, which deviates from the normal axis of cross section. Therefore the curvature of centerline is determined not only by the rigid rotation of cross section, as well as the shear deformation. The internal force and torque are proportional to the strains and curvature-twist respectively. The dynamical equations of the Cosserat's rod can be established with Cardan's angles ψ,ϑ,φ and deflections w_i $(i=1,2,3)$ as unknown variables. The configuration of the rod after deformation is determined by adding the deflections to the initial shape.

As an example we consider the motion of an axial moving beam with constant velocity v_0 and axial extensive force F_0 (Fig.10). Since the mass of beam flows along the centerline, the Euler's concept of velocity field is applied in mathematic formulation. As a special case of Cosserat's model with small deformation, the dynamical equation of an axial moving Timoshenko's beam can be written directly

$$a_1\frac{\partial^4 w}{\partial t^4}+a_2\frac{\partial^4 w}{\partial t^3\partial s}+a_3\frac{\partial^4 w}{\partial t^2\partial s^2}+a_4\frac{\partial^2 w}{\partial t^2}+a_5\frac{\partial^4 w}{\partial t\partial s^3}+a_6\frac{\partial^2 w}{\partial t\partial s}+a_7\frac{\partial^4 w}{\partial s^4}+a_8\frac{\partial^2 w}{\partial s^2}=0 \tag{6.1}$$

where the complex variable $w=w_1+\mathrm{i}w_2$ is applied to describe the three-dimensional deformation of the beam. In the stability analysis of quasi-stationary state of the beam the critical axial velocity before buckling can be derived as

$$v_{0,\mathrm{cr}} = \left[\frac{F_0}{\rho S}\left(\frac{1-\alpha\sigma_n}{1-\sigma_n}\right)\right]^{1/2}, \quad \alpha = \frac{\kappa GS}{F_0}, \sigma_n = \frac{EIn^2\pi^2}{F_0 l^2(1-\alpha)} \quad (n=1,2,\cdots) \tag{6.2}$$

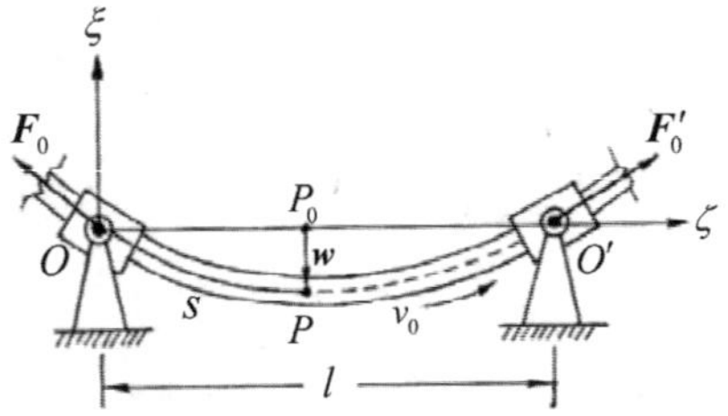

Figure 10. The axial moving beam

References

[1] Liu YZ, Xue Y 2004 *Technische Mechanik* **24** 206~210
[2] Xue Y, Liu YZ, Chen LQ 2005 *Chinese J. Theor.& Appl. Mech.* **37** 485-493 (in Chinese)
[3] Xue Y, Liu YZ, Chen LQ 2006 *Acta Physica Sinica* , **55** 3845~3851 (in Chinese)
[4] Xue Y, Wang P 2011 *Acta Physica Sinica* **60** 114501-5 (in Chinese)
[5] Wang P, Xue Y, Liu YL 2012 C*hin. Physics* B**21** 070203
[6] Xue Y, Shang HL 2009 *Chinese Physics Letters*, **26** 074501-3， /7
[7] Liu YZ 2006 *Nonlinear mechanics of thin elastic rod* (Beijing Tsinghua University Press & Springer) (in Chinese)
[8] Liu YZ, Zu JW 2004 *Acta Mechanica* **167** 29-39
[9] Liu YZ 2003 *Chinese Quarterly of Mechanics* **24** 433-439 (in Chinese).
[10] Liu YZ, Xue Y, Chen LQ 2004 *Acta Physica Sinica* **53** 2424-2428 (in Chinese).
[11] Xue Y, Chen LQ Liu YZ 2004 *Acta Physica Sinica* **53** 4029-4035 (in Chinese).
[12] Liu YZ, Xue Y 2005 *Chinese Quarterly of Mechanics* **26** 1-7 (in Chinese).
[13] Liu YZ 2005 *Acta Mechanica Solida Sinica* **26** 256-260 (in Chinese).
[14] Peng JH, Liu YZ 2005 *J. Dynamics and Control* **3** 36-39 (in Chinese)
[15] Liu YZ, Sheng LW 2007 *Acta Physica Sinica* **56** 2305-2310 (in Chinese).
[16] Liu YZ, Sheng LW 2007 *Chinese Physics* **16** 891-896
[17] Liu YZ, Sheng LW 2007 *Acta Mechanica Sinica* **23** 215-219
[18] Liu YZ， Xue Y 2009 *Acta Physica Sinica* **58** 5936-5941(in Chinese)
[19] Xue Y，Weng DW Gong GM 2010 *First International Conference on Cellular, Molecular Biology, Biophysics and Bioengineering/Volume* IV/300/2010
[20] Xue Y, Weng DW, Gong GM 2012 *Advanced Materials Research* **343-344** 661-667
[21] Liu YZ, Xue Y 2005 *J. Applied Mechanics*, **22** 392-394 (in Chinese)
[22] Liu YZ, Xue Y 2008 *Chinese Quarterly of Mechanics* **29** 343-348 (in Chinese)
[23] Liu YZ 2011 *Chinese Quarterly of Mechanics* **32** 295-299 (in Chinese)
[24] Liu YZ, Xue Y 2011 *Acta Mechanica Sinica* **43** 1151-1156 (in Chinese)
[25] Liu YZ 2009 *Chinese Physics* B**18** 1-8
[26] Liu YZ, Xue Y 2011 *Applied Mathematics & Mechanics* **32** 603-612
[27] Liu YZ 2012 *Acta Mechanica Sinica* **44** 832-838 (in Chinese)
[28] Xue Y, Weng DW, Chen LQ 2013 *Acta Physics Sinica* **62** 0446011-0446017 (in Chinese)

High-order Vibration Characteristics of Rotating Thin Shells and Hard-coating Damping Effects

X Y Song [1], H J Ren [1], X P Wang [2], X J Li [3] and Q K Han [1,3*]

[1] School of Mechanical Engineering, Dalian University of Technology, Dalian, PR China 116024
[2] Shenyang Engine Design and Research Institute, Shenyang, PR China, 110015
[3] Hunan Provincial Key Laboratory of Health Maintenance for Mechanical Equipment, Hunan University of Science and Technology, Xiangtan, PR China, 411201

E-mail: hanqingkai@dlut.edu.cn

Abstract. The high-order modes and the moving wave characteristics of rotating thin shell are investigated in this paper based on transfer matrix method when the damping effects of the hard-coating layers are applied on the shell. Firstly, the dynamic model of the thin shell is obtained based on the Love theory, with introducing centrifugal force and Coriolis force. Then, the transfer matrix formulas of the shell with partially coated layers at different segments are obtained under the clamped-free boundary condition, in which the mechanical property of the hard-coating material is simplified to be linear and isotropic. The modal characteristics of the shell are calculated numerically. The obtained higher order modal frequencies of the partially coated shells are compared with those of the bare shell. The effects of the material parameters and locations of the hard-coating layers are also illustrated based on the different modal characteristics., great care should be taken in constructing both.

1. Introduction

Rotating thin shells are widely used in many industry areas including the shafts of turbine machines, machining tools, combined disc-drum, etc. Thin shells often possess advantages of good stiffness and low mass. However, they are also prone to suffer from serious vibration and high cycle fatigue (HCF) due to multi-harmonic resonance, rubbing-impact and highly rotating speed. In practical design, the lower order resonances of a rotating shell is not permitted and must be guaranteed to avoid firstly. Besides, the higher-order vibrations of the thin shell also need to be paid attention because of their direct contribution to HCF. Sometimes, special damping treatments are urgently developed to compensate the lifetime of the shell under certain operating condition.

Nowadays, there are many classic theories and methods to analyze the vibrations of rotating thin shells. The natural characteristics of a rotating ring were investigated and moving waves were observed at the first time by Bryan [1]. The Coriolis effect on rotating shell was discussed by Taranto and Lessen [2]. The natural characteristics and the forward/backward moving waves of a rotating cylindrical shell with infinite length were investigated for different Coriolis and centrifugal forces by

[1,3] hanqingkai@dlut.edu.cn

Srinivasan and Lauterbach [3]. While the rotating shell with finite length were discussed by Zohar and Aboudi[4], Wang and Chen [5].

The vibrations of shells with different boundary conditions and corresponding solving methods were also exploredextensively. The natural frequencies of the clamp-clamp supported by rotating shell were solved based on triangle-Galerkin combined method by Saito and Endo[6]. The natural frequencies of rotating cylindrical shell with different boundary conditions were solved using complex exponent method by Penzes and Kraus [7]. The forward and backward moving waves were compared by Lam and Loy[8], for both different boundary conditions (simple-simple, clamped-clamped, clamped-free, respectively) and different shell theories (Donnell, Flügge, Love and Sander's theory). The effects of the rotating speed and thickness were also reported by them [9].

Among the solving methods of shell dynamics, the transfer matrix method was introduced to solve the free vibration of clamp-free boundary condition shell by Tottenham and Shimizn [10]. The dynamic characteristics of conical and cylindrical shells were solved using transfer matrix method by Irie [11]. The forward and backward moving waves of the rotating shell were compared by Hong, Guo and Zhu [12-13], for different thicknesses and Coriolis effects. The transfer matrix method can solve complex shell, the high order vibration character of the rotating thin shell were solved by Han [14], considering the effect of sealing teeth.

Besides, constrained layer damping (CLD) is widely used in the aerospace and automotive industries as a novel technique of surface treatment to attenuate resonant structural vibrations of shells. The deformation energy of the structure is dissipated through the compliant layer made of viscoelastic material, sandwiched between the base structure and a stiff constraining layer. The performances of such treatment have been studied by many researchers. The overall damping achieved has been shown to be dependent not only on the inherent damping in the viscoelastic material but also on the thickness and elastic modulus of each layer [15].

Hard-coating treatment is well known to play an important role in improving the structure lifetime and performance [16]. In the last two decades, a so-called hard-coating damping technology for structure vibration reduction is interesting for many people. It can be achieved by using special coating materials with porous micro-structures that couple mechanical behaviour with electrical or magnetic fields applied to them [17].

Most of the previous research, however, for a rotating thin shell, the effect of hard-coating damping and different coating parts, have no analytical, numerical and experiment results at all. This paper presents a strategy to solve the high-order vibrations of rotating thin shell based on transfer matrix method, and the damping effects of hard-coating treatment on its natural characteristics are compared. The method described in this paper is believed to include several novelties: First, the vibration equation of the rotating shell is established, considering the effect of centrifugal and Coriolis forces. Furthermore, based on Love shell model the governing differential equations for cylindrical shell segments are derived. Then, the state space vectors, propagation matrices and precise integrations are formed for different segments with or without hard-coating layers along axial directions. The mechanical property of the hard-coating material is simplified to be linear and isotropic. The obtained natural frequencies, especially higher-order modes, of the shell with different hard-coating segments are compared finally.

2. Analytic model of the rotating cylindrical shell and transfer matrix equations

Figure 1 shows a partly cylinder treated with hard-coating damping material which rotates in angular velocity speed Ω. The shell is modelled as a cylindrical shell consisting of sealing teeth effects. The figure 1(a) is a model of the partly coating damping material shell and the figure 1(b) is the cutaway view of the shell. The circular cylindrical shell is with a radius of R, thickness H and length L. The curve coordinate system is $o\text{-}x\theta z$. The variables u, v and w denotes the displacement of the shell in x-axis, y-axis and z-axis. The shell is divided into N parts and the lengths of every parts are L_1, $L_2, \cdots L_N$. The damping marital length is $b = x_2 - x_1$, where x_2 and x_1 denote the two ends of the damping parts in x direction.

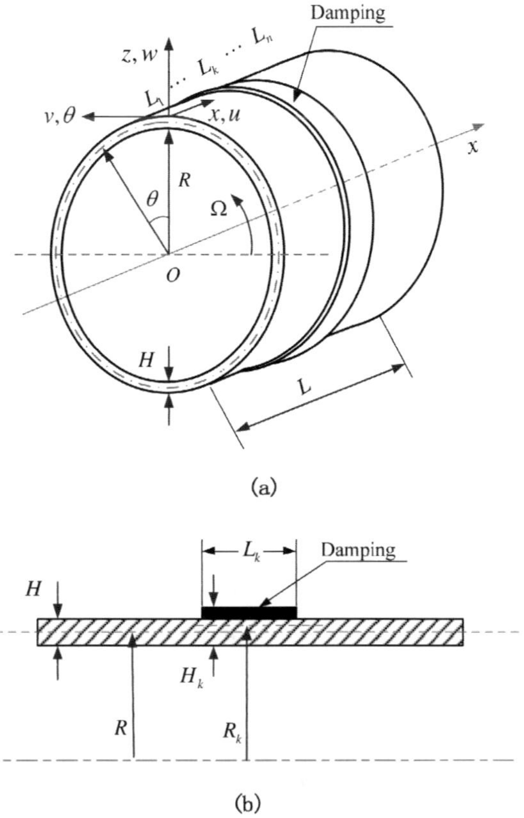

Figure 1. Rotating cylindrical shell partly treated with hard-coating damping material and its cutaway view.

The strains of the neutral surface of the shell, $\varepsilon_x^{(0)}$, $\varepsilon_\theta^{(0)}$ and shearing strain $\gamma_{x\theta}^{(0)}$, can be written as follows

$$\varepsilon_x^{(0)} = \frac{\partial u}{\partial x},\ \varepsilon_\theta^{(0)} = \frac{\partial v}{R\partial\theta},\gamma_{x\theta}^{(0)} = \frac{\partial v}{\partial x} + \frac{\partial u}{R\partial\theta} \tag{1}$$

Based on the Love thin shell theory, the intersection angle of the shell θ_x and θ_θ are described as

$$\theta_x = -\frac{\partial w}{\partial x},\ \theta_\theta = \frac{v}{R} - \frac{\partial w}{R\partial\theta} \tag{2}$$

The curvature change of the neutral surface κ_x , κ_θ and torsion $\chi_{x\theta}$ are

$$\kappa_x = -\frac{\partial^2 w}{\partial x^2},\ \kappa_\theta = \frac{1}{R^2}\left[\frac{\partial v}{\partial\theta} - \frac{\partial^2 w}{\partial\theta^2}\right],\ \chi_{x\theta} = \frac{1}{R}\left[\frac{\partial v}{\partial\theta} - 2\frac{\partial^2 w}{\partial x\partial\theta}\right] \tag{3}$$

For a thin shell, the displacements in x and θ directions are assumed to change linearly with the thickness, and the displacements in the transverse direction are independent on z. Thus, the strain-displacement relations based on Love theory are, therefore, with the following forms

$$\varepsilon_x = \varepsilon_x^{(0)} + z\kappa_x,\ \varepsilon_\theta = \varepsilon_\theta^{(0)} + z\kappa_\theta,\ \gamma_{x\theta} = \gamma_{x\theta}^{(0)} + z\chi_{x\theta} \tag{4}$$

where, z is the displacement of a point in transverse direction.

Based on the Hooke's law

$$\sigma_x = \frac{E}{1-\mu^2}(\varepsilon_x + \mu\varepsilon_\theta), \sigma_\theta = \frac{E}{1-\mu^2}(\varepsilon_\theta + \mu\varepsilon_x), \tau_{x\theta} = \frac{E}{2(1+\mu)\gamma_{x\theta}} \tag{5}$$

where, σ_x is the stress in x-axis, σ_θ is the stress in θ-axis and $\tau_{x\theta}$ is the shear force in x-θ plane.

The internal force is expressed as follow

$$N_x = \int_{-H/2}^{H} \sigma_x \mathrm{d}z, N_\theta = \int_{-H/2}^{H} \sigma_\theta \mathrm{d}z, N_{x\theta} = \int_{-H/2}^{H} \sigma_{x\theta} \mathrm{d}z \tag{6a}$$

$$M_x = \int_{-H/2}^{H/2} \sigma_x z\mathrm{d}z, M_\theta = \int_{-H/2}^{H/2} \sigma_\theta z\mathrm{d}z, M_{x\theta} = \int_{-H/2}^{H/2} \sigma_{x\theta} z\mathrm{d}z \tag{6b}$$

where, N_x, N_θ and $N_{x\theta}$ are the shear forces of the neutral surface per unit length, M_x, M_θ and $M_{x\theta}$ are the bending moment and torsion moment of the neutral surface per unit length.

Substituting equation (4) and (5) into equation (6) yields

$$N_x = K\left[\frac{\partial u}{\partial x} + \frac{\mu}{R}(\frac{\partial v}{\partial \theta} + w)\right] \tag{7a}$$

$$N_\theta = K\left[\frac{1}{R}(\frac{\partial v}{\partial \theta} + w) + \mu\frac{\partial u}{\partial x}\right] \tag{7b}$$

$$N_{x\theta} = K\frac{1-\mu}{2}\left(\frac{\partial v}{\partial x} + \frac{1}{R}\frac{\partial u}{\partial \theta}\right) \tag{7c}$$

$$M_x = D\left[-\frac{\partial^2 w}{\partial x^2} + \frac{\mu}{R^2}\left(\frac{\partial v}{\partial \theta} - \frac{\partial^2 w}{\partial \theta^2}\right)\right] \tag{7d}$$

$$M_\theta = D\left[\frac{1}{R^2}\left(\frac{\partial v}{\partial \theta} - \frac{\partial^2 w}{\partial \theta^2}\right) - \mu\frac{\partial^2 w}{\partial x^2}\right] \tag{7e}$$

$$M_{x\theta} = D\frac{1-\mu}{2R}\left(\frac{\partial v}{\partial x} - 2\frac{\partial^2 w}{\partial x \partial \theta}\right) \tag{7f}$$

where, K is film stiffness and D is bending rigidity

$$K = \frac{EH}{1-\mu^2}, D = \frac{EH^3}{12(1-\mu^2)}$$

The linear differential equation of the rotating thin shell can be written as follow. Equation (8a) is the oscillatory differential equation of the axial vibration. Equation (8b) is the oscillatory differential equation of the circumferential vibration and equation (8c) is the oscillatory differential equation of the radial vibration.

$$\frac{\partial N_x}{\partial x} + \frac{1}{R}\frac{\partial N_{x\theta}}{\partial \theta} + N_\theta^0\left(\frac{1}{R^2}\frac{\partial^2 u}{\partial \theta^2} - \frac{1}{R}\frac{\partial w}{\partial x}\right) = \rho H\frac{\partial^2 u}{\partial t^2} \tag{8a}$$

$$\frac{1}{R}\frac{\partial N_\theta}{\partial \theta} + \frac{\partial N_{x\theta}}{\partial x} + \frac{Q_\theta}{R} + \frac{N_\theta^0}{R}\frac{\partial^2 u}{\partial x \partial \theta} = \rho H\frac{\partial^2 v}{\partial t^2} + 2\rho H\Omega\frac{\partial w}{\partial t} - \rho H\Omega^2 v \tag{8b}$$

$$\frac{\partial Q_x}{\partial x} + \frac{\partial Q_\theta}{R\partial \theta} - \frac{N_\theta}{R} + \frac{N_\theta^0}{R^2}\left(\frac{\partial^2 w}{\partial \theta^2} - \frac{\partial v}{\partial \theta}\right) = \rho H\frac{\partial^2 w}{\partial t^2} - 2\rho H\Omega\frac{\partial v}{\partial t} - \rho H\Omega^2 w \tag{8c}$$

where, N_θ^0 is the strain due to the centrifugal force; $N_\theta^0 = \rho H\Omega^2 R^2$; $\rho H\frac{\partial^2 u}{\partial t^2}, \rho H\frac{\partial^2 v}{\partial t^2}, \rho H\frac{\partial^2 w}{\partial t^2}$ are inertia force; $2\rho H\Omega\frac{\partial w}{\partial t}$, $2\rho H\Omega\frac{\partial v}{\partial t}$ are Coriolis force and $\rho H\Omega^2 v$, $\rho H\Omega^2 w$ are centrifugal force. Q_x and Q_θ are shear force per unit length in x and θ direction, and can be written as follows

$$Q_x = \frac{\partial M_x}{\partial x} + \frac{\partial M_{x\theta}}{R\partial\theta} \tag{9a}$$

$$Q_\theta = \frac{\partial M_{x\theta}}{\partial x} + \frac{\partial M_\theta}{R\partial\theta} \tag{9b}$$

The general solution of equation (8) is expressed as follows

$$u(x,\theta,t) = \sum_{m=0}^{\infty}\sum_{n=0}^{\infty} \tilde{u}(x)\cos(n\theta \pm \omega t) \tag{10a}$$

$$v(x,\theta,t) = \sum_{m=0}^{\infty}\sum_{n=0}^{\infty} \tilde{v}(x)\sin(n\theta \pm \omega t) \tag{10b}$$

$$w(x,\theta,t) = \sum_{m=0}^{\infty}\sum_{n=0}^{\infty} \tilde{w}(x)\cos(n\theta \pm \omega t) \tag{10c}$$

where, ω is the natural frequency of the shell; m,n are longitudinal half-wave number and the circumferential wave number; '+'and'−' is the forward traveling wave and backward wave of the rotating shell.

Substituting Equation (10) into Equation (2) yields

$$\theta_x = \tilde{\theta}_x \cos(n\theta \pm \omega t) \tag{11a}$$

$$\theta_\theta = \tilde{\theta}_\theta \sin(n\theta \pm \omega t) \tag{11b}$$

where, $\tilde{\theta}_x = -d\tilde{w}/dx$ and $\tilde{\theta}_\theta = \tilde{v}/R + n\tilde{w}/R$.

Substituting Equation (10) into Equation (9a) and Equation (9b), one obtains that that

$$\begin{bmatrix} N_x & N_\theta & M_x & M_\theta & Q_x & V_x \end{bmatrix}^T = \begin{bmatrix} \tilde{N}_x & \tilde{N}_\theta & \tilde{M}_x & \tilde{M}_\theta & \tilde{Q}_x & \tilde{V}_x \end{bmatrix}^T \cos(n\theta \pm \omega t) \tag{12a}$$

$$\begin{bmatrix} N_{x\theta} & M_{x\theta} & Q_\theta & S_x \end{bmatrix}^T = \begin{bmatrix} \tilde{N}_{x\theta} & \tilde{M}_{x\theta} & \tilde{Q}_\theta & \tilde{S}_x \end{bmatrix}^T \sin(n\theta \pm \omega t) \tag{12b}$$

where, $\tilde{N}_x = K\left[\frac{d\tilde{u}}{dx} + \mu\left(\frac{n}{R}\tilde{v} + \frac{1}{R}\tilde{w}\right)\right]$,

$\tilde{M}_x = D\left(\frac{d\tilde{\theta}_x}{dx} + \mu\frac{n}{R}\tilde{\theta}_\theta\right)$, $\tilde{N}_\theta = K\left(\frac{n}{R}\tilde{v} + \frac{1}{R}\tilde{w} + \mu\frac{d\tilde{u}}{dx}\right)$, $\tilde{M}_\theta = D\left(\frac{n}{R}\tilde{\theta}_\theta + \mu\frac{d\tilde{\theta}_x}{dx}\right)$,

$\tilde{N}_{x\theta} = K\frac{1-\mu}{2}\left(\frac{d\tilde{v}}{dx} - \frac{n}{R}\tilde{u}\right)$, $\tilde{M}_{x\theta} = D\frac{1-\mu}{2}\left(\frac{d\tilde{\theta}_\theta}{dx} - \frac{n}{R}\tilde{\theta}_x\right)$, $\tilde{Q}_x = \frac{d\tilde{M}_x}{dx} + \frac{n}{R}\tilde{M}_{x\theta}$, $\tilde{Q}_\theta = \frac{d\tilde{M}_{x\theta}}{dx} - \frac{n}{R}\tilde{M}_\theta$,

$\tilde{V}_x = \tilde{Q}_x + \frac{n}{R}\tilde{M}_{x\theta}$, $\tilde{S}_x = \tilde{N}_{x\theta} + \frac{1}{R}\tilde{M}_{x\theta}$.

Substituting equation (10) and equation (11) into vibration equation (8) yields

$$\frac{d\tilde{N}_x}{dx} + \frac{n}{R}\tilde{N}_{x\theta} - \rho H\Omega^2 R\left(\frac{n^2}{R}\tilde{u} + \frac{d\tilde{w}}{dx}\right) + \rho H\omega^2\tilde{u} = 0 \tag{13a}$$

$$\frac{d\tilde{N}_{x\theta}}{dx} - \frac{n}{R}\tilde{N}_\theta + \frac{1}{R}\tilde{Q}_\theta - n\rho H\Omega^2 R\frac{d\tilde{u}}{dx} + \rho H\left(\omega^2\tilde{v} \pm 2\omega\Omega\tilde{w} + \Omega^2\tilde{v}\right) = 0 \tag{13b}$$

$$\frac{d\tilde{Q}_x}{dx} + \frac{n}{R}\tilde{Q}_\theta - \frac{\tilde{N}_\theta}{R} - \rho H\Omega^2\left(n^2\tilde{w} + n\tilde{v}\right) + \rho H\left(\omega^2\tilde{w} \pm 2\omega\Omega\tilde{v} + \Omega^2\tilde{w}\right) = 0 \tag{13c}$$

From the second factor and third factor of equation (13), the control equation contain the terms of $\pm 2\omega\Omega$. It is why the rotating shell has two different frequencies which are defined as the forward traveling wave and backward wave.

The primary variables of the rotating cylindrical shell are expressed in the vector form as

$$\tilde{Z}(x)=[\tilde{u},\tilde{v},\tilde{w},\tilde{\theta}_x,\tilde{M}_x,\tilde{V}_x,\tilde{S}_x,\tilde{N}_x]^T \tag{14}$$

The general solution of the state variables is expressed as

$$Z(x,\theta,t)=\sum_{m=0}^{\infty}\sum_{n=0}^{\infty}\begin{bmatrix}\tilde{u}(x)\cos(n\theta\pm\omega t)\\ \tilde{v}(x)\sin(n\theta\pm\omega t)\\ \tilde{w}(x)\cos(n\theta\pm\omega t)\\ \tilde{\theta}_x(x)\cos(n\theta\pm\omega t)\\ \tilde{M}_x(x)\cos(n\theta\pm\omega t)\\ \tilde{V}_x(x)\cos(n\theta\pm\omega t)\\ \tilde{S}_x(x)\sin(n\theta\pm\omega t)\\ \tilde{N}_{\mathrm{x}}(x)\cos(n\theta\pm\omega t)\end{bmatrix} \tag{15}$$

where ω is the natural frequency of the rotating shell.

Based on equation (11)- equation (13), the only state equation for the primary variables is considered for simplification

$$\frac{\mathrm{d}\tilde{Z}(x)}{\mathrm{d}x}=\tilde{U}\cdot\tilde{Z}(x) \tag{16}$$

where, $\tilde{U}$ is a constant coefficient matrix of 8×8 and $\tilde{U}$ is expressed as follows

$$\tilde{U}=\begin{bmatrix}0 & U_{12} & U_{13} & 0 & 0 & 0 & 0 & U_{18}\\ U_{21} & 0 & 0 & U_{24} & 0 & 0 & U_{27} & 0\\ 0 & 0 & 0 & U_{34} & 0 & 0 & 0 & 0\\ 0 & U_{42} & U_{43} & 0 & U_{45} & 0 & 0 & 0\\ U_{51} & 0 & 0 & U_{54} & 0 & U_{56} & U_{57} & 0\\ 0 & U_{62} & U_{63} & 0 & U_{65} & 0 & 0 & U_{68}\\ 0 & U_{72} & U_{73} & 0 & U_{75} & 0 & 0 & U_{78}\\ U_{81} & 0 & 0 & U_{84} & 0 & 0 & U_{87} & 0\end{bmatrix} \tag{17}$$

The elements which is not equal zero in matrix $U_{ij}(i,j=1,\cdots,8)$ are relate to the natural frequency ω of the shell, rotating speed Ω, geometrical parameter and material parameter. The expressions of the matrix $\tilde{U}$ are given in Appendix A.

The rotating shell in figure1 is divided into n_0 parts. The propagating relationship given in equation (16) can be used repeatedly so that we can propagate the physical quantities from the one edge to the other edge of the rotating shell. Thus, we have

$$\tilde{Z}_{n_0}(\xi_{n_0})=\tilde{T}(\omega)\cdot\tilde{Z}_1(\xi_0) \qquad (i=1,...,n_0) \tag{18}$$

Where, $\tilde{T}(\omega)$ is the propagator matrix, which can be written as $\tilde{T}(\omega)=\prod_{i=1}^{n_0}\tilde{T}_i(\omega)$.

According to the differential equations theory, state vector equations can be expressed as

$$\tilde{Z}(\xi_i)=\tilde{T}_i(\omega)\,\tilde{Z}(\xi_{i-1}) \quad (i=1,...,n_0) \tag{19}$$

where, for the ith rotating cylindrical thin shell, the propagator matrix can be written as $\tilde{T}_i(\omega)=\exp(\tilde{U}_iL_i)$.

In order to get a more exact result, the ith shell can be divided into k parts, whose size is L_1, L_2, ..., L_{k-1}, L_k in $x-$direction, and the transfer matrix is defined as follow

$$\tilde{Z}(L_k)=\tilde{G}(L_k)\tilde{Z}(L_{k-1}) \tag{20}$$

where, $\tilde{G}(L_k)$ is the transfer matrix of the kth shell and $\tilde{G}(L_k)=\exp(\tilde{U}_k L_k)$.

When the character of a rotating thin shell is solved, the characteristic equation is defined by boundary condition.

For a clamp-free supported rotating cylindrical thin shell, the boundary conditions are as follows:

$u = v = w = \varphi_x = 0$ where $x = 0$,

$\tilde{M}_x = \tilde{V}_x = \tilde{S}_x = \tilde{N}_x = 0$ where $x = L$.

Substituting the boundary conditions into equation (14), then yields

$$\begin{bmatrix} 0 \\ 0 \\ 0 \\ 0 \end{bmatrix} = \begin{bmatrix} \tilde{T}_{11} & \tilde{T}_{12} & \tilde{T}_{13} & \tilde{T}_{14} \\ \tilde{T}_{21} & \tilde{T}_{22} & \tilde{T}_{23} & \tilde{T}_{24} \\ \tilde{T}_{31} & \tilde{T}_{32} & \tilde{T}_{33} & \tilde{T}_{34} \\ \tilde{T}_{41} & \tilde{T}_{42} & \tilde{T}_{43} & \tilde{T}_{44} \end{bmatrix} \begin{bmatrix} \tilde{u}_x \\ \tilde{v}_x \\ \tilde{w}_x \\ \tilde{\varphi}_x \end{bmatrix} \tag{21}$$

The hard-coating damping material is anisotropic. In this paper, the Voigt approach method is introduced to solve the hard-coating damping shell.

Here E_d is Young's modulus of the hard damping material; E_m is Young's modulus of the thin shell; V_d is the volume fraction of the hard damping material; V_m is the volume fraction of thin shell; μ_d is Poisson's ratio of damping material and μ_m is Poisson's ratio of shell, based on the Voigt approach method then yields

$$E = E_d V_d + E_m V_m \tag{22}$$

$$\mu = \mu_d V_d + \mu_m \mu_m \tag{23}$$

$$G = \frac{G_d G_m}{G_d V_d + G_m V_m} \tag{24}$$

$$\rho = \rho_d V_d + \rho_m V_m \tag{25}$$

3. Numerical results

The geometrical parameter and material parameter of the shell is given in table1.

Table 1. Properties and geometrical parameter of the thin cylindrical shell

Properties	Symbol	Value	Units
Young's modulus	E	206×10^{11}	Pa
Poisson's ratio	μ	0.3	
Density	ρ	7850	kg/m^3
Length	L	0.256	m
Radius	R	0.141	m
Thickness	H	0.002	m

The hard-coating thin shell composed of an isotropic metal substrate and a hard-coating damping layer is analyzed. The hard-coating damping layer is made of Al2O3+MgO, which is an anisotropic material. The material parameters of this kind of hard-coating materials of Al_2O_3+MgO are listed in table 2.

Table 2. Material parameters of hard-coating damping material

Elastic coefficients C_{ik} Unit Gpa								
C11	C12	C13	C22	C23	C33	C44	C55	C66
183.1	173.2	170.5	183.1	170.5	183.1	45.3	45.3	56.5

In this paper we set the Young's modules $E_d = 170.5\,\text{GPa}$, material density $\rho_d = 5300\,\text{kgm}^{-3}$ and Poisson's ratio $\mu_d = 0.3$. Based on equation (22) and equation (25), the linear Young's modules are calculated as $E = 204\times10^9\,\text{Pa}$, density $\rho = 7660\,\text{kgm}^{-3}$. Substituting the linear material parameters into equation (18), the natural frequency of the partially hard-coating damping material shell can be calculated.

3.1 The high-order vibration characteristics of the bare shell when rotating speed $\Omega = 0$

First of all, the transfer matrix method and finite element method are introduced to solve the static shell with the clamp-free boundary condition. For boundary conditions, the natural frequencies can be solved from equation (21). It is assumed that the top and bottom edges are clamp and free boundaries.

In order to validate the transfer matrix method results, the finite element method (FEM) is used to calculate the natural frequency values of the corresponding mode orders for the non-coated shell under the same boundary condition. The finite element model of one of the thin shell here composes of 920 elements (SHELL181 of ANSYS) of the substrate thin shell. The natural frequencies and mode shapes are solved by Block Lanczos method.

Figure 2 shows the natural frequency of the static shell with different half-wave number m and the circumferential wave number n. The curve $m = 1$ shows that the transfer matrix method and FEM have the same precision. The frequencies based on the transfer matrix corresponding to modes 1–5 are exactly the same as that by FEM. The natural frequency of the thin shell decreases firstly, then enlarge from 4389.4 Hz to 4864.4 Hz, at the (1,6) step the natural is the lowest 1355.5 Hz. Compare with $m = 1$ and $m = 2$ the natural frequency is different with the same circumferential wave number n.

When $n < 6$, the difference between the transfer matrix method and the FEM is slight, but the difference increases with the circumferential wave number changes from 1 to 20.

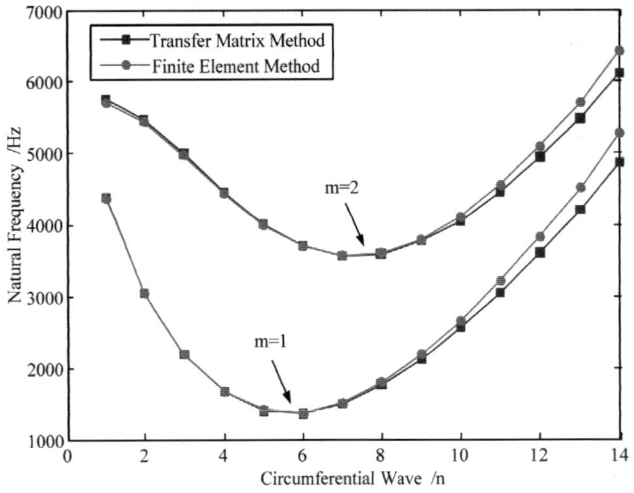

Figure 2. The high-order frequencies of the static shell.

3.2 The Forward and backward waves of the high-order modes of the bare shell under different rotating speeds

Figure 3 and figure 4 show the relationship between the rotating speed Ω and the natural frequency. Based on equation (12)- (18) the transfer matrix of a clamp-free boundary condition shell with rotating speed Ω. When the rotating speed Ω increases, the natural frequency of the rotating shell are divided into two parts, the top curve is the forward traveling wave frequency and the other one is the backward traveling wave. Different from the low-order condition, the high-order vibration's backward frequency curve increase with the rotating speed changed from 0 rad/sec to 2000rad/sec. In figure 8 and figure 10 the difference between FEM and TRM are enlarged as the rotating speed increases.

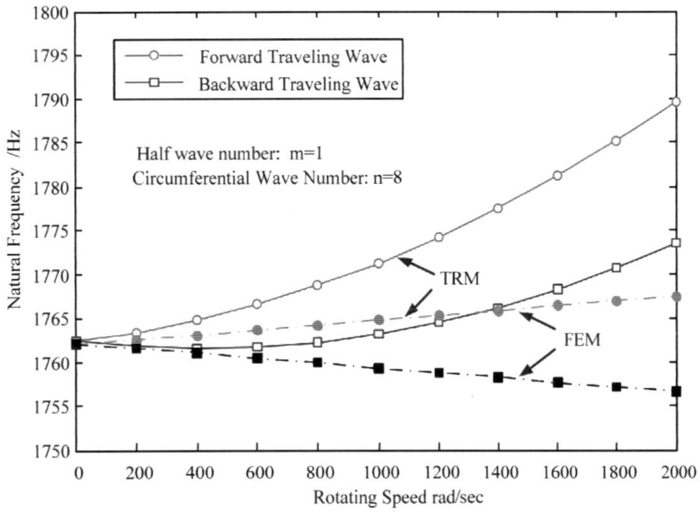

Figure 3. Forward and backward waves of the rotating cylindrical shells with clamped-free supported boundary conditions at both edges for high nodal diameters n=8.

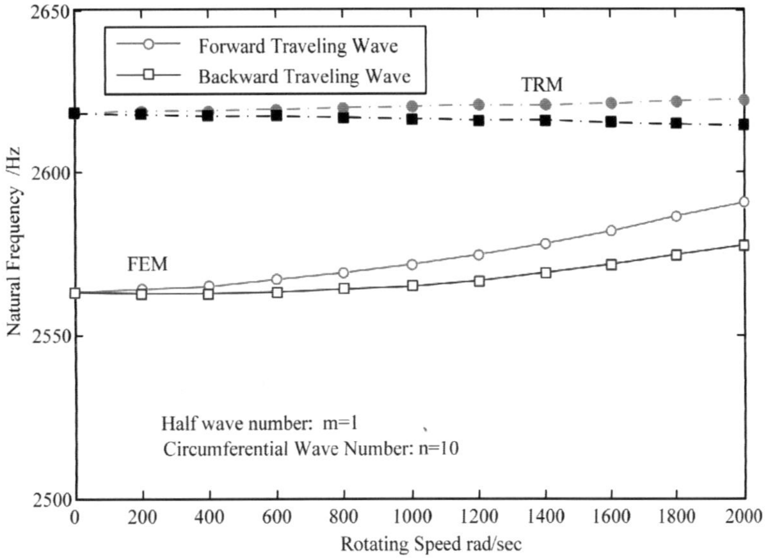

Figure 4. Forward and backward waves of the rotating cylindrical shells with clamped-free supported boundary conditions at both edges for high nodal diameters n=10.

3.3 The high-order vibration characteristics of the hard-coating shell when rotating speed $\Omega = 0$

The hard-coating layers are applied on the total out surface of the shell by APS. Figure 5 compares the natural frequency of hard-coating shell and thin shell. By checking the obtained modes of the bare plate and the coated shell with hard-coating damping material, it is observed that the natural frequency changes. When the half-wave number *m*=1, the first 6 natural frequencies of the two shells values have a very little difference. When the mode changes from *n*=7 to *n*=20, the different between the shell and hard-coating damping shell increases at the same, especially at the high-order vibration.

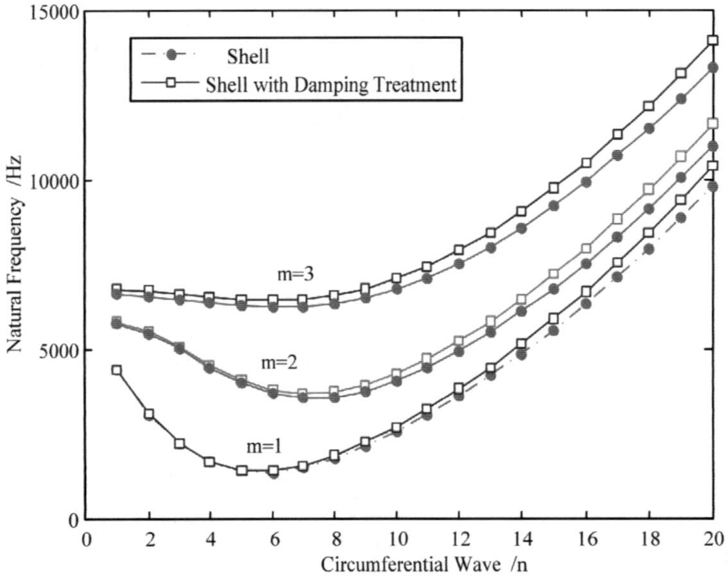

Figure 5. The natural frequencies of the shell with hard-coating damping material treatment totally.

3.4 The high-order vibration characteristics of the partly coated shell on the middle position when rotating speed

The transfer matrix method is introduced to solve the partially hard-coating damping material shell which is 30% area of the total shell, shown in figure 1, with the clamp-free boundary condition. In this paper, the coating position is in the middle of the *x*-axis and the results are shown in figure 6. By checking the obtained modes of the bare shell and the coated shell with 30% hard-coating damping material, it is obtained that the natural frequency has the same changing trend. In the case $m = 1, n \leq 13$, the coating damping material of 30% area has little effect on the thin shell; at the high vibration order $m = 1, n > 13$, the damping material has an obviously effect on the natural frequency. In the case $m > 1$, the damping material works better than that when *m*=1.

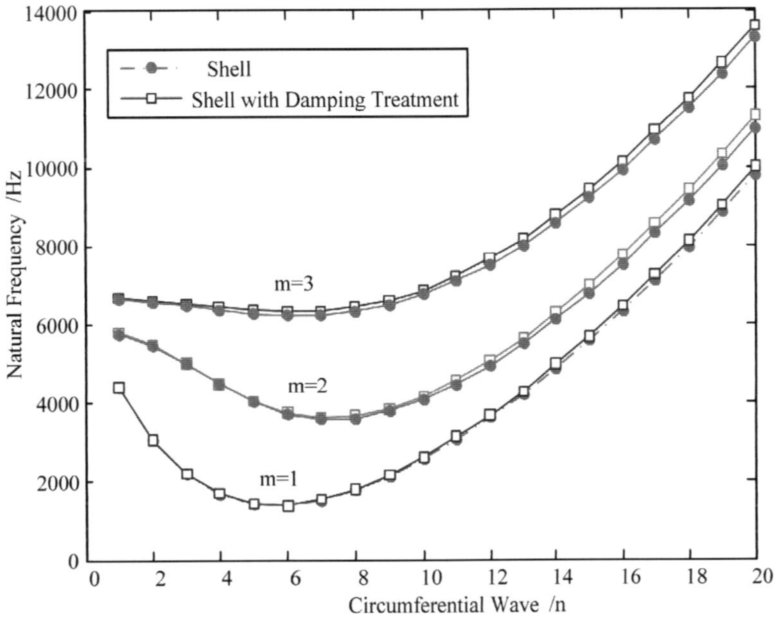

Figure 6. The natural frequencies of the shell with hard-coating damping material treatment of 30% area.

3.5 The high-order vibration characteristics of the partly coated shells on differnt positions when rotating speed $\Omega=0$

In order to compare different cover position of the damping material effect on the shell, the results are shown in table 2 and figure 6. It is clearly that the natural frequencies of the shells are different due to the effect of the cover position of the hard-coating damping material. When the material is covered near the free edge, the natural frequency of the shell increases more obviously than the other two cases. The natural frequencies have the same trend due to the circumferential wave number *n*.

Table 3. Comparison of different coating position for the natural frequency of the shell (the coating area is 30%)

Order	Coating position			Order	Coating position		
	Top	Middle	Bottom		Top	Middle	Bottom
1	4386.4	4405.8	4424.5	**11**	3185.6	3119.3	3085.4
2	3040.3	3063.1	3085.6	**12**	3759.3	3680.8	3639.6
3	2189.1	2209.3	2229.4	**13**	4386.5	4295.2	4246.4
4	1673.2	1688.6	1704.7	**14**	5066.2	4960.9	4904.3
5	1412.8	1420.2	1430.4	**15**	5797.1	5677.3	5612.7
6	1379.2	1374.8	1377.2	**16**	6579.2	6443.9	6371.1
7	1534.7	1516.8	1511.1	**17**	7412.2	7260.3	7179.1
8	1825.8	1794.9	1781.7	**18**	8295.7	8126.5	8036.7
9	2211.1	2168.1	2147.9	**19**	9229.6	9042.1	8943.8
10	2667.9	2613.3	2586.3	**20**	10213.6	10007.7	9900.3

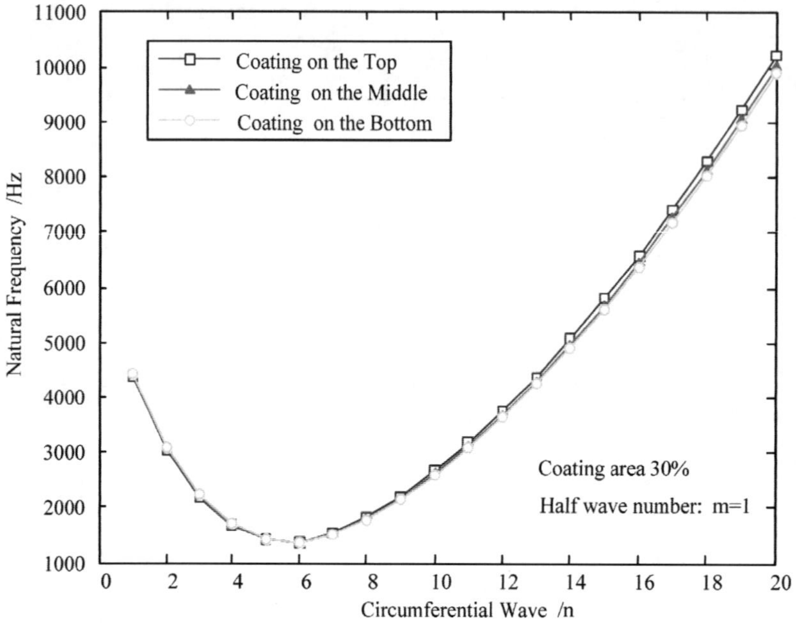

Figure 7. Comparison of different coating positions of the natural frequencies of the shell.

3.6 The high-order vibration characteristics of the partly coated shells on middle position with different rotating speeds

Figure 8 shows the relationship between the rotating speed Ω and the natural frequency of the hard-coating damping material. While the rotating speed Ω becomes fast, the natural frequency of the rotating shell are divided into two parts, the top curve is the forward traveling wave frequency and the other one is the backward traveling wave. In this figure, we consider the case that the half wave number m=1 and circumferential wave number n=10. At the high rotating speed, the backward traveling frequency is near to the forward traveling frequency in this case. The damping material has little effect on the rotating shell.

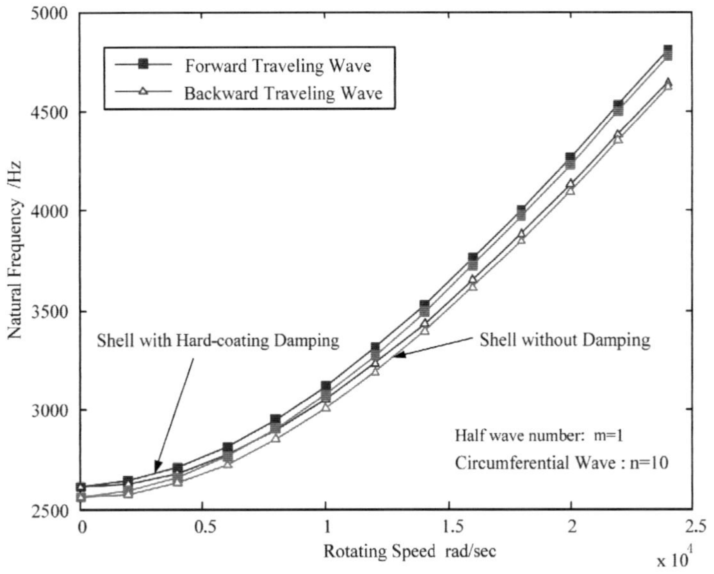

Figure 8. Bifurcations of the natural frequency of hard-rotating damping cylindrical shells with coating area of 30% for high nodal diameters.

4. Conclusions

This paper presents a transfer matrix method to analyze the high-order vibrations and the damping effects of the hard-coating layers applying on different ring-like areas. The conclusions are conducted as follows:

(1) The transfer matrix method used for the bare shell under the same boundary condition has the same precision compare with the finite element method. And the transfer also calculates the natural frequency of the rotating shell.

(2) The Voigt approach method is introduced to solve the hard-coating damping shell and the anisotropic material can be approximated by linear material. The Voigt approach can calculate the hard-coating damping material shell.

(3) By checking the obtained modes of the bare shell and the coated shell with hard-coating damping material, it is observed that the natural frequency have the same change trend. At the high vibration order, the damping material has a more obviously effect on the natural frequency than at the low order.

(4) The covering position of the damping material has different effect on the natural frequency of the thin shell. The natural frequency values of the shell increase more obviously than the clamp and middle edges.

Reference

[1] Bryan G H 1890 On the beats in the vibrations of a revolving cylinder or bell *Proceedings of the Cambridge Philosophical Society* **7** 100-101.

[2] Lessen R A M 1964 Coriolis acceleration effect on the vibration of a rotating thin walled circular cylinder *Journal of Applied Mechanics* **31** 700–701.

[3] Srinivasan A V, Lauterbach G F 1971 Traveling waves in rotating cylindrical shells *Journal of Engineering for Industry, ASME* **93** 1229-1232.

[4] Zohar, Aboudi J 1973 The free vibrations of thin circular finite rotating cylinder *International Journal of Mechanical Sciences* **15** 269–278.

[5] Wang S S, Chen Y 1974 Effects of rotation on vibrations of circular cylindrical shells *Journal of the Acoustical Society of America* **55** 1340–1342.
[6] Saito T, Endo M. 1986 Vibration of finite length, rotating cylindrical shells *Journal of Sound and Vibration* **107** 17-28.
[7] Penzes L E, Kraus H 1972 Free vibrations of prestressed cylindrical shells having arbitrary homogeneous boundary conditions *AIAA* **10** 1309-1313.
[8] Lam K Y, Loy C T 1995 Analysis of rotating laminated cylindrical shells by different thin shell theories *Journal of Sound and Vibration,* **186** 23-35.
[9] Lam K Y, Loy C T 1998 Influence of boundary conditions for a thin laminated rotating cylindrical shell *Composite Structures,* **41** 215-228.
[10] Tottenham H, Shimizu K 1972 Analysis of the free vibration of cantilever cylindrical thin elastic shells by the matrix progression method *International Journal of Mechanical Sciences* **14** 293-310.
[11] Irie T, Yamada G and Kaneko Y 1982 Free vibration of a conical shell with variable thickness *Journal of Sound and Vibration* **82** 83-94.
[12] Hong J, Guo B T and Zhu Z G 1999 The research of rotating thin shell and the experiment *Aircraft Engine* **1** 40-44.
[13] Hong J, Guo B T and Zhu Z G 1998 The experiment research of rotating thin shell *Journal of Aerospace Power* **13** 40-44.
[14] Han Q K, Wang Y 2012 The high order vibration character of the rotating thin shell with the effect of sealing teeth *SCIENCE CHINA Physics,Mechanics & Astronomy.*
[15] Rongong J A, Tomlinson G R 1996 Suppression of ring vibration modes of high nodal diameter using constrained layer damping methods *Smart Mater. Struct.* **5** 672-684.
[16] Ding J N, Meng Y G and Wen S Z 2000 Mechanical properties and fracture toughness of multilayer hard coatings using nanoindentation *Thin Solid Films* **371** 178-182.
[17] Buravalla V R, Remillat C and Rongong J A 2001 Advances in damping materials and technology *Smart Materials Bulletin* **8** 10-13.

Appendix

$$U_{12}=-\frac{\mu n}{R},\ U_{13}=-\frac{\mu}{R},\ U_{18}=\frac{1-\mu^2}{EH},\ U_{21}=\frac{12nR}{H^2-12R^2},\ U_{24}=\frac{2H^2n}{H^2+12R^2},$$

$$U_{27}=\frac{24R^2(1+\mu)}{EH(H^2+12R^2)},\ U_{34}=-1,\ U_{42}=-\frac{\mu n}{R^2},\ U_{43}=-\frac{\mu n^2}{R^2},\ U_{45}=\frac{12(1+\mu^2)}{EH^3},$$

$$U_{51}=-\frac{EH^3n^2}{R(1+\mu)(H^2+12R^2)},\ U_{54}=\frac{2EH^3n^2}{(1+\mu)(H^2+12R^2)},\ U_{56}=1,\ U_{57}=-\frac{2H^2n}{H^2+12R^2},$$

$$U_{62}=\rho H\left(n\Omega^2\pm\omega\Omega\right)+n\frac{EH}{R^2}\left(1+\frac{n^2H^2}{12R^2}\right),\ U_{63}=-\rho H\left[\omega^2-(m^2-1)\Omega^2\right]+\frac{EH}{R^2}(\frac{H^2n^4}{12R^2}+1),$$

$$U_{65}=\frac{\mu n^2}{R^2},\ U_{68}=\frac{\mu}{R},\ U_{72}=-\rho H\left[\omega^2+(1+\mu n^2)\Omega^2\right]+n^2\frac{EH}{R^2}\left(1+\frac{H^2}{12R^2}\right),\ U_{75}=\frac{\mu n}{R^2}$$

$$U_{73}=-\rho H\left(\pm 2\omega\Omega+\Omega^2\right)+n\frac{EH}{R^2}\left(1+\frac{n^2H^2}{12R^2}\right),\ U_{78}=\frac{\mu n}{R}+\frac{n(1-\mu^2)\rho R\Omega^2}{E},$$

$$U_{81}=-\rho H\left(\omega^2-n\Omega^2\right)+\frac{n^2EH^3}{2R^2(1+\mu)(H^2+12R^2)},$$

$$U_{84}=-\rho H\Omega^2R-\frac{n^2EH^3}{R(1+\mu)(H^2+12R^2)},\ U_{87}=-\frac{12nR}{H^2+12R^2}.$$

Dynamic analysis of slender launching system connected by clamp band joint using harmonic balance method

Z Y Qin[1], S Z Yan and F L Chu

Department of Mechanical Engineering, Tsinghua University, China

qinzy@mail.tsinghua.edu.cn

Abstract. Clamp band joints are widely used to fasten spacecrafts onto launching systems. Due to the unilateral constraints and the frictional slippage at the joint interface, clamp band joints may bring nonlinearity into launching systems during launching process. In this paper, the dynamics of a slender launching system with clamp band joint is investigated using harmonic balance method. Firstly, the formulas for the joint stiffness of the clamp band joint are proposed. Then, the finite element model for the launch vehicle and the spacecraft connected by the clamp band joint is developed, where the clamp band joint is represented by a massless beam element. Finally, harmonic balance method is applied to calculate the steady state response of the launching system.

1. Introduction

Spacecrafts are commonly fastened onto launch vehicles by the clamp band joints in modern aerospace applications. The continuity of the interface between the launch vehicle and the spacecraft is guaranteed by the preload imposed onto the clamp band joint. Though the magnitude of the preload is strictly controlled to prevent joint separation, there may exist frictional slippage at the joint interface under severe loading conditions, which results in variations of the local stiffness and in turn leads to nonlinear dynamic behavior of the launching system.

Dynamics of launch vehicles have been extensively investigated. For different research purposes, various dynamic models for launch vehicles incorporating the propulsion [1], aerodynamics [2], guidance and control [3], and structural flexibility [4] were developed. When modeling launching systems, the approximation of rigid connection between launch vehicles and spacecrafts is usually adopted without accommodating the influence of clamp band joint.

The authors have developed the finite element model for a launching system, where the clamp band joint is represented by a massless beam element [5]. The modal characteristics of the launching system were discussed in detail considering the variations of the mass and length of the launching system due to the fuel combustion and stage jettisons during the ascent flight; whereas, the steady state response was simply calculated for certain excitation frequencies by numerical integration.

Compared with numerical integration, harmonic balance method (HBM) is a more efficient approach and has been widely used for solving strongly nonlinear problems [6-9].

[1] Corresponding author: Department of Mechanical Engineering, Tsinghua University, Haidian District, Beijing, China, 100084. email: qinzyl@mail.tsinghua.edu.cn

In this paper, the launching system model accommodating the nonlinear influence of the clamp band joint proposed in [5] is adopted. HBM is applied to computing the steady state response of the launching system and evaluating the influence of the clamp band joint on the system response.

2. Modelling of Launching System with Clamp Band Joint

2.1. Clamp Band Joint Model

The configuration of the clamp band joint is shown in Figure 1. Under preload, the joint generates axial and bending constraints and fastens the interface ring and the payload adapter together, which connect to the spacecraft and launch vehicle respectively. The transverse deformation at the joint interface is resisted by auxiliaries such as lips, pins or splines. Hence, the joint is assumed to be rigid in the lateral direction.

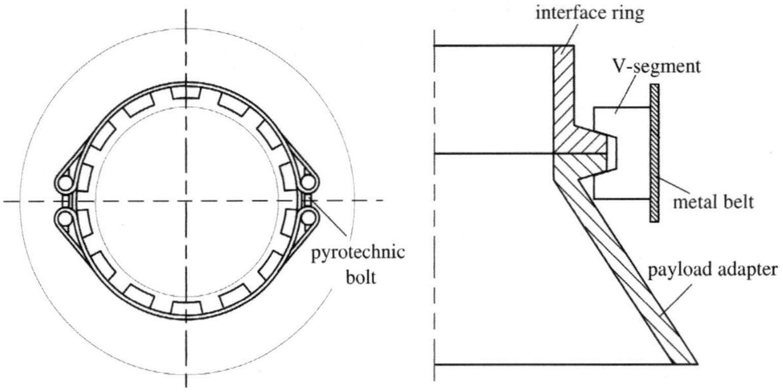

Figure 1. Sketch of the joint interface between the launch vehicle and the spacecraft.

The relative axial deformation at the joint interface under the axial load T is given by [5]

$$Y=\frac{\frac{r_0T}{\pi R_0}-2\left(R_1-R_{f0}\right)f_v+\left(t_r-\frac{t_{f0}}{2}\right)\frac{S_0}{R_b}}{\frac{k_r}{\Delta R}+k_p+\frac{EI_r}{\overline{R}^2\Delta R}} \tag{1}$$

where, $r_0=R_1-R_0-\frac{\nu R_0A\left(\lambda t_0+1\right)}{2\lambda^2R_0{}^2A+2\lambda\overline{R}^2t_s}$, $k_r=\frac{EAt_s\left(2+\lambda t_0+\frac{\overline{R}^2t_s}{\lambda R_0{}^2A}\right)}{4\lambda^3R_0{}^2A+4\lambda^2\overline{R}^2t_s}$, $t_r=\frac{\overline{R}^2t_s\left(\lambda t_0+1\right)}{2\lambda^2R_0{}^2A+2\lambda\overline{R}^2t_s}$,

$$f_v=\frac{(\cos\varphi-\mu\sin\varphi)S_0}{2(\sin\varphi+\mu\cos\varphi)R_b}+\frac{\mu S_0}{\left(\sin^2\varphi-\mu^2\cos^2\varphi\right)R_b}\frac{T}{T_{\mathrm{m}}},\ k_p=\begin{cases}0, & Y>0\\ \frac{E\left(R_1-R_3\right)^3\left(R_1+3R_3\right)}{6t_0\overline{R}\Delta R}, & Y\le 0\end{cases}.$$

The relative deflection angle at the joint interface under the bending moment M is [5]

$$\Psi_R=\frac{R_1-R_3}{R_3}\left(\psi^0-\psi^\pi\right) \tag{2}$$

where,

$$\psi^0=\frac{\left[\frac{t_r}{2}-\frac{(\cos\varphi-\mu\sin\varphi)(R_1-R_{f0})}{2(\sin\varphi+\mu\cos\varphi)}-\frac{t_{f0}}{4}\right]\frac{S}{R_b}}{k_r+\frac{EI_r}{\overline{R}^2}}+\frac{\left[\frac{r_0}{\pi R_0^2}-\frac{(R_1-R_{f0})\mu S}{\left(\sin^2\varphi-\mu^2\cos^2\varphi\right)R_bM_{\mathrm{m}}}\right]M}{k_r+\frac{EI_r}{k_m\overline{R}^2}},$$

$$\psi^{\pi} = \begin{cases} \dfrac{\left[\dfrac{t_r}{2} - \dfrac{(\cos\varphi - \mu\sin\varphi)(R_1 - R_{f0})}{2(\sin\varphi + \mu\cos\varphi)} - \dfrac{t_{f0}}{4}\right]\dfrac{S}{R_b}}{k_r + \dfrac{EI_r}{\bar{R}^2}} - \dfrac{\left[\dfrac{r_0}{\pi R_0^2} + \dfrac{(R_1 - R_{f_0})\mu S}{2(\sin^2\varphi + \mu\sin\varphi\cos\varphi)R_b M_{\mathrm{m}}}\right]M}{k_r + \dfrac{EI_r}{k_m \bar{R}^2}}, & \psi^{\pi} > 0 \\ 0, & \psi^{\pi} \le 0 \end{cases}.$$

Correspondingly, the axial stiffness K_a and bending stiffness K_m of the clamp band joint can be expressed in the incremental form

$$K_a = \frac{\delta T}{\delta Y} \tag{3}$$

$$K_m = \frac{R_3 \delta M}{(R_1 - R_3)(\delta\psi^0 - \delta\psi^{\pi})} \tag{4}$$

The relationships between the external loads and relative deformations at the joint interface calculated based on Eqs. (3) and (4) are shown in Figure 2, where the structural parameters of the 1194 adapter of LM-3A and the preload of 25 kN are adopted. It is revealed in Fig. 2 that the both the relation curves are piecewise linear.

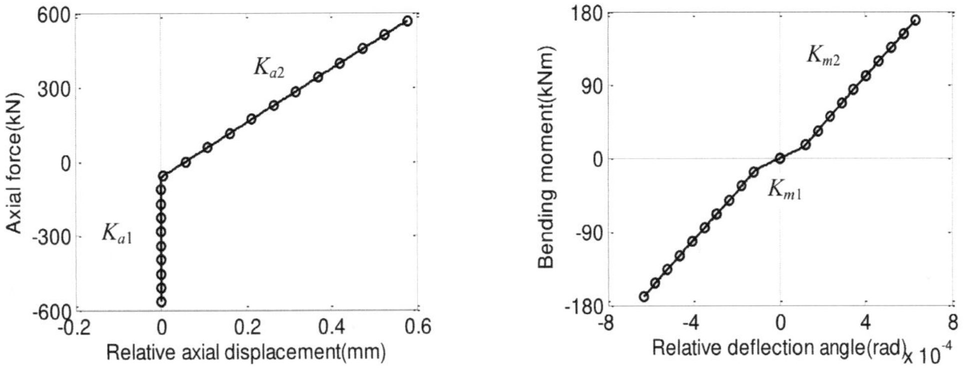

(a) Axial force vs. axial deformation (b) Bending moment vs. deflection angle

Figure 2. Relationship between external loads and relative deformation at the joint interface.

2.2. Launching System Model

The launching system is divided into five components, that is, the 1st, 2nd and 3rd stages of the launch vehicle, the payload fairing and the spacecraft. Those components are modeled separately using the beam element shown in Fig. 3. By assuming that the launch system is axisymmetric and neglecting its torsional vibration, the motion of launching system is in the xy plane. Correspondingly, there are three degrees of freedom (dofs) in each node, and the vector of generalized displacements of the beam element is

$$\boldsymbol{\delta}^{\mathrm{e}} = \{u_1 \quad v_1 \quad \theta_1 \quad u_2 \quad v_2 \quad \theta_2\}^{\mathrm{T}} \tag{5}$$

The mass and stiffness matrices of the beam element are developed based upon Timoshenko beam theory. The propellant in the launch vehicle is modeled as additional mass without rotary inertia and imposed to the relevant nodes of the beam models to represent the influence of the liquid on the structural dynamics. The element mass matrix of the liquid propellant can be written as

$$\mathbf{M}_{\mathrm{e}}^{\mathrm{L}} = \mathrm{diag}\begin{bmatrix} \dfrac{m^l}{2n} & \dfrac{m^l}{2n} & 0 & \dfrac{m^l}{2n} & \dfrac{m^l}{2n} & 0 \end{bmatrix} \tag{6}$$

Then the element mass matrix for each stage of the launch vehicle is

$$\mathbf{M}_{\mathrm{e}} = \mathbf{M}_{\mathrm{e}}^{\mathrm{S}} + \mathbf{M}_{\mathrm{e}}^{\mathrm{L}} \tag{7}$$

By assembling the component models together, the linear model for the launching system without considering the influence of the clamp band joint can be obtained as

$$\mathbf{M}\ddot{\boldsymbol{\delta}} + \mathbf{C}\dot{\boldsymbol{\delta}} + \mathbf{K}\boldsymbol{\delta} = \mathbf{F}(\mathbf{t}) \tag{8}$$

The clamp band joint connecting the launch vehicle and the spacecraft is treated as a massless beam element as shown in Fig. 4, where the joint stiffness matrix $\mathbf{K_{cb}}$ involves the axial and bending stiffnesses as bellow

$$\mathbf{K_{cb}} = \begin{bmatrix} K_a & 0 \\ 0 & K_m \end{bmatrix} \tag{9}$$

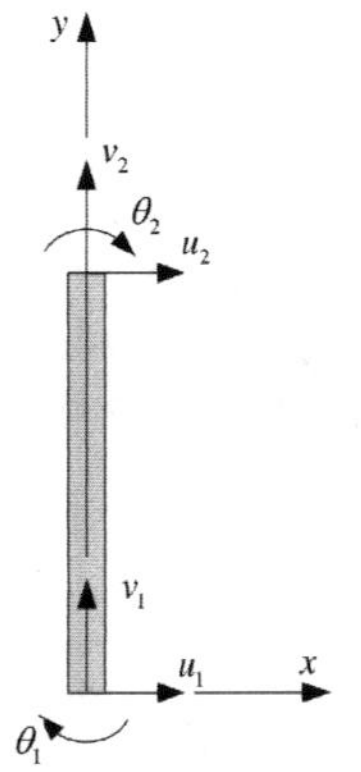

Figure 3. Beam element of launching system.

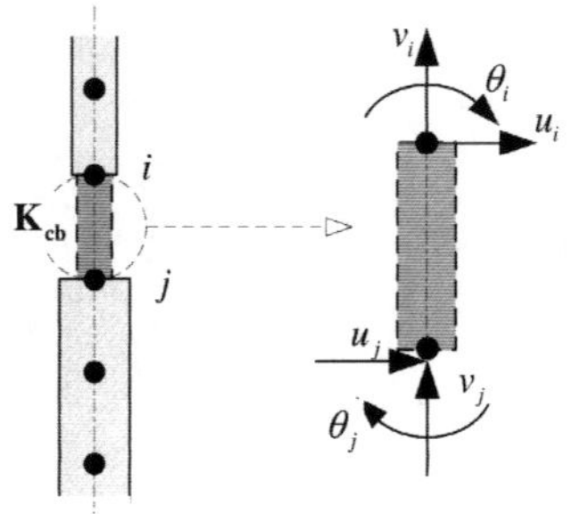

Figure 4. Connection between the beam models of launch vehicle and spacecraft.

Neglecting the joint mass, the general form of equation for the launching system accommodating the clamp band joint is

$$\mathbf{M}\ddot{\boldsymbol{\delta}} + \mathbf{C}\dot{\boldsymbol{\delta}} + \mathbf{K}\boldsymbol{\delta} = \mathbf{F}(\mathbf{t}) - \mathbf{K_J}\boldsymbol{\delta} \tag{10}$$

where, $\mathbf{K_J}$ is the joint stiffness matrix extended from $\mathbf{K_{cb}}$, whose expression is

$$\mathbf{K_J} = \begin{array}{c} \\ 1 \\ \vdots \\ i \\ \vdots \\ \vdots \\ j \\ \vdots \\ \vdots \\ \vdots \end{array} \begin{array}{c} \begin{array}{cccccccc} 1 & \cdots & i & \cdots & \cdots & j & \cdots & \cdots \end{array} \\ \begin{bmatrix} \mathbf{0} & & & & & & & \\ & \ddots & & & & & & \\ & & \mathbf{K_{cb}} & & & \mathbf{-K_{cb}} & & \\ & & & \mathbf{0} & & & & \\ & & & & \ddots & & & \\ & & \mathbf{-K_{cb}} & & & \mathbf{K_{cb}} & & \\ & & & & & & \mathbf{0} & \\ & & & & & & & \ddots \\ & & & & & & & & \mathbf{0} \end{bmatrix} \end{array} \tag{11}$$

The damping matrix of the launching system model is built according to Rayleigh damping law

$$\mathbf{C} = \alpha\mathbf{M} + \beta\mathbf{K} \tag{12}$$

where the constants α and β are such as the modal damping for the first and six modes of the system are equal to 0.4%.

For convenience of description, the system model given in Eq. (8) is referred to as the fixed model; whereas, that given by Eq. (10) is referred to as the clamped model.

3. Harmonic Balance Method with Condensation

3.1. Harmonic Balance Formulation

HBM is employed to calculate the steady state response of the launching system. Considering the response to be periodic, the displacements $\boldsymbol{\delta}(t)$ can be represented as a truncated Fourier series with N harmonics

$$\boldsymbol{\delta}(t)=\tilde{\mathbf{a}}_0+\sum_{n=1}^{N}\left(\tilde{\mathbf{a}}_n\cos n\Omega t+\tilde{\mathbf{b}}_n\sin n\Omega t\right) \tag{13}$$

Denote $\mathbf{R}_{\mathbf{cb}}=\mathbf{K}_{\mathbf{cb}}\boldsymbol{\delta}$, which represents the nonlinear force caused by the clamp band joint. Correspondingly, the nonlinear force $\mathbf{R}_{\mathbf{cb}}(t,\boldsymbol{\delta})$ can also be written as a truncated Fourier series

$$\mathbf{R}_{\mathbf{cb}}(t,\boldsymbol{\delta})=\tilde{\mathbf{c}}_0+\sum_{n=1}^{N}\left(\tilde{\mathbf{c}}_n\cos n\Omega t+\tilde{\mathbf{d}}_n\sin n\Omega t\right) \tag{14}$$

Substituting the expressions of $\boldsymbol{\delta}(t)$ and $\mathbf{R}_{\mathbf{cb}}(t,\boldsymbol{\delta})$ from Eqs. (13) and (14) into Eq. (10) and balancing the harmonic terms yields

$$\boldsymbol{\Lambda}\mathbf{Z}+\mathbf{d}-\mathbf{F}=\mathbf{0} \tag{15}$$

where, $\mathbf{Z}=\left[\tilde{\mathbf{a}}_0^{\mathrm{T}},\ \tilde{\mathbf{a}}_1^{\mathrm{T}},\ \tilde{\mathbf{b}}_1^{\mathrm{T}},\ \cdots,\ \tilde{\mathbf{a}}_N^{\mathrm{T}},\ \tilde{\mathbf{b}}_N^{\mathrm{T}}\right]^{\mathrm{T}}$, $\mathbf{d}=\left[\tilde{\mathbf{c}}_0^{\mathrm{T}},\ \tilde{\mathbf{c}}_1^{\mathrm{T}},\ \tilde{\mathbf{d}}_1^{\mathrm{T}},\ \cdots,\ \tilde{\mathbf{c}}_N^{\mathrm{T}},\ \tilde{\mathbf{d}}_N^{\mathrm{T}}\right]^{\mathrm{T}}$ and $\mathbf{F}$ are $4(m+2)(2N+1)\times 1$ vectors of Fourier coefficients of the displacement and nonlinear force and excitation force, $\boldsymbol{\Lambda}$ is a block-diagonal matrix of $4(m+2)(2N+1)\times 4(m+2)(2N+1)$ dimension, whose expression is

$$\boldsymbol{\Lambda}=\begin{bmatrix}\mathbf{K} & & & & & \\ & \boldsymbol{\Lambda}_1 & & & & \\ & & \ddots & & & \\ & & & \boldsymbol{\Lambda}_n & & \\ & & & & \ddots & \\ & & & & & \boldsymbol{\Lambda}_N\end{bmatrix} \tag{16}$$

where

$$\boldsymbol{\Lambda}_k=\begin{bmatrix}\mathbf{K}-(k\Omega)^2\mathbf{M} & k\Omega(\mathbf{C}+\Omega\mathbf{G})\\ -k\Omega(\mathbf{C}+\Omega\mathbf{G}) & \mathbf{K}-(k\Omega)^2\mathbf{M}\end{bmatrix} \tag{17}$$

The Fourier coefficients $\mathbf{d}$ are nonlinear functions of $\mathbf{Z}$ in an implicit form in Eq. (15), where the relationship between $\mathbf{d}$ and $\mathbf{Z}$ can be obtained by using the alternate frequency/time domain (AFT) method [10] sketched as follow:

$$\mathbf{Z}\xrightarrow{\text{IDFT}}\boldsymbol{\delta}(t)\xrightarrow{\mathbf{K}_{\mathbf{be}}}\mathbf{R}_{\mathbf{be}}(t,\boldsymbol{\delta})\xrightarrow{\text{DFT}}\mathbf{d}$$

Then only $4(m+2)(2N+1)$ unknowns of $\mathbf{Z}$ elements left in Eq. (15). And those simultaneous algebraic equations can be solved by using the Netwon-Raphson method iteratively, given a proper guess of the initial values.

3.2. Condensation

The finite element model for the launching system can contain quite a large number of dofs, which makes the solving procedure time consuming. On the contrary, the nonlinear influence of the clamp band joint is only related to the axial and deflection dofs at the joint interface. Therefore, condensation is carried out to reducing the problem order.

The equations in Eq. (15) are reorganized into p linear dofs and q nonlinear dofs for each harmonic as follow

$$\begin{bmatrix}\boldsymbol{\Lambda}_{qq} & \boldsymbol{\Lambda}_{qp}\\ \boldsymbol{\Lambda}_{pq} & \boldsymbol{\Lambda}_{pp}\end{bmatrix}\begin{bmatrix}\mathbf{Z}_q\\ \mathbf{Z}_p\end{bmatrix}=\begin{bmatrix}\mathbf{F}_q\\ \mathbf{F}_p\end{bmatrix}-\begin{bmatrix}\mathbf{d}_q\\ \mathbf{d}_p\end{bmatrix} \tag{18}$$

From Eq. (18), the Fourier coefficients $\mathbf{Z}_q$ of the displacement of the nonlinear dofs can be expressed as

$$\hat{\mathbf{\Lambda}}_{qq}\mathbf{Z}_q = \hat{\mathbf{F}}_q - \mathbf{d}_q \tag{19}$$

where, $\hat{\mathbf{\Lambda}}_{qq} = \mathbf{\Lambda}_{qq} - \mathbf{\Lambda}_{qp}\mathbf{\Lambda}_{pp}^{-1}\mathbf{\Lambda}_{pq}$, $\hat{\mathbf{F}}_q = \mathbf{F}_q - \mathbf{\Lambda}_{qp}\mathbf{\Lambda}_{pp}^{-1}\mathbf{F}_p$. The Netwon-Raphson method can be applied to solving Eq. (19) to obtain $\mathbf{Z}_q$ with a huge saving in computational cost. Consequently, the Fourier coefficients $\mathbf{Z}_p$ of the linear dofs can be obtained using the relation

$$\mathbf{Z}_p = \mathbf{\Lambda}_{pp}^{-1}\left(\mathbf{F}_p - \mathbf{\Lambda}_{pq}\mathbf{Z}_q\right) \tag{20}$$

4. Computations and Results

4.1. Model Parameters

The three-stage expendable LM-3A is studied here, of which the configuration is shown in Fig. 5. The LM-3A is a liquid fueled launch vehicle, of which the mass values of the three stages and payload faring are listed in Table 1 [11]. The spacecraft mounted on the launch vehicle is assumed to be a beam with equivalent length of 3.5 m and diameter of 0.4 m. When performing response analyses, both lateral and longitudinal excitations are imposed at the bottom of the launching system model.

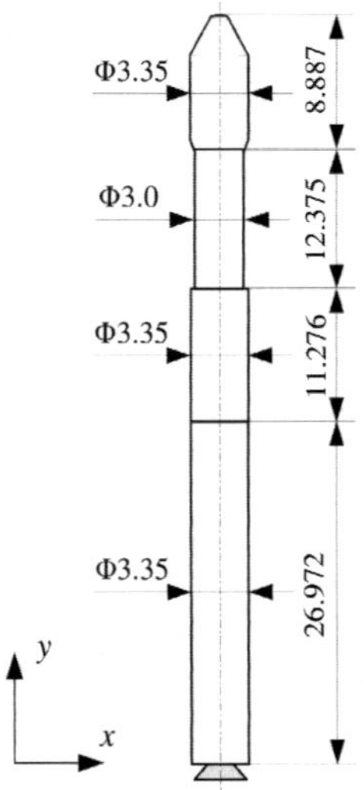

Figure 5. Schematic configuration of LM-3A launch vehicle (Units: m).

Table 1. Parameter of LM-3A launch vehicle (Units: kg).

Component	Structural mass	Propellant mass
1st stage	11.23×103	172.05×103
2nd stage	3.561×103	30.555×103
3rd stage	2.742×103	18.193×103
Payload fairing	500	—

4.2. Validation of Harmonic Balance Method

Prior to investigating the steady state response of the launching system using HBM, numerical integration is undertaken to validate the HBM. A harmonic excitation with the amplitude of 1g and the frequency of 15 Hz is applied to the launching system model in the longitudinal direction. The displacement response at the top of the spacecraft and the relative displacement at the joint interface are computed by numerical integration and HBM are illustrated in Fig. 6, where harmonic terms of 3, 7 and 9 are considered for the HBM.

It can be seen from Fig. 6(a) that the spacecraft response calculated from HBM with different retained harmonic terms shows good correlation with that of the numerical integration. However, obvious divergence from the numerical integration results can be observed for the case of $n = 3$ when examining the relative displacement at the joint interface shown in Fig. 6(b). Seven harmonic terms provide sufficient accuracy with sound agreement between the simulation results of the HBM and those of the numerical integration. Hence, $n = 7$ is adopted in the following analyses.

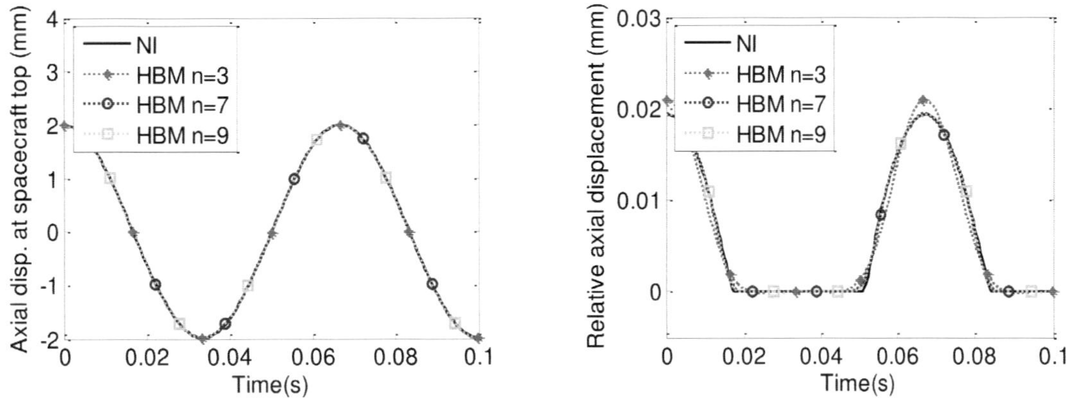

(a) Displacement at top of spacecraft (b) Relative displacement at joint interface

Figure 6. Comparison of simulation results of numerical integration and HBM.

4.3. *Steady State Response of Launching System*

The frequency response curve of the launching system subjected to the lateral excitation with the amplitude of 50 kN and frequencies ranging from 1-50 Hz is shown in Fig. 7. It is revealed from Fig. 7 that for relatively lower frequencies the results calculated from the clamped model and the fixed one show little difference. As the excitation frequency increases, the response amplitude of the clamped model becomes slightly larger than that of the fixed model; whereas, along with the further increment of the excitation frequency, the response amplitude of clamped model gets smaller compared with that of the fixed model.

The response amplitudes at the top of the spacecraft under the longitudinal excitation of 1g in the frequency band of 1-50 Hz is illustrated in Fig. 8, where no significant difference between the simulation results of clamped and fixed models is observed. Though there exists relative displacement between the launch vehicle and the spacecraft in the clamped model, the relative displacement is so small compared with the response that it has little effect on the launching system response.

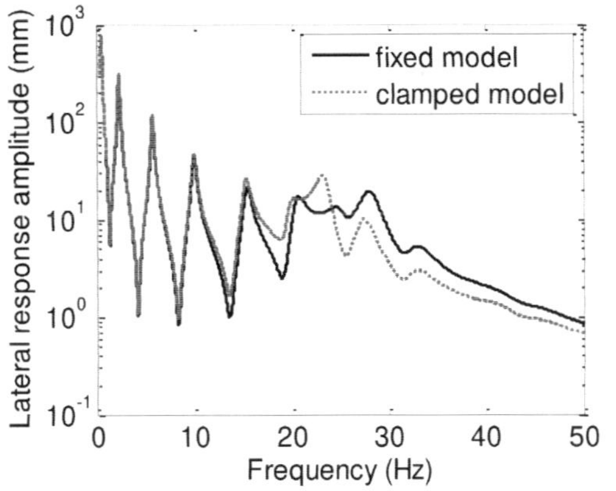

Figure 7. Frequency response curve of launching system at top of spacecraft in lateral direction.

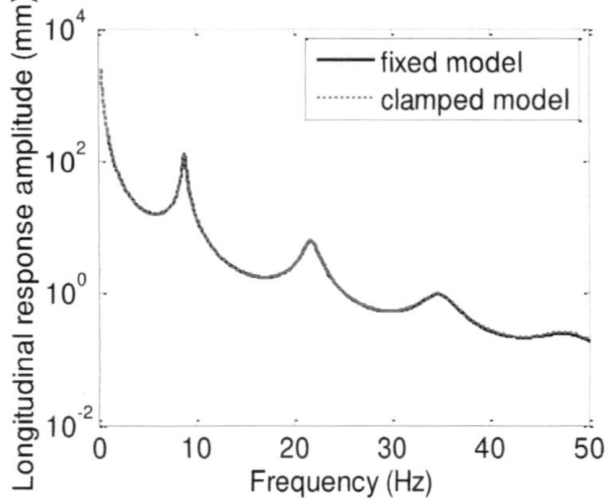

Figure 8. Frequency response curve of launching system at top of spacecraft in the longitudinal direction.

5. Conclusions

HBM was employed to calculate the steady state response of the launching system, where the clamp band joint is modelled as a massless beam element with nonlinear stiffnesses in both the lateral and longitudinal directions. It was revealed that the nonlinear joint stiffnesses had little influence on the longitudinal response of the launching system and affected the lateral response for relatively high frequency band.

Actually, there do exist relative displacements at the joint interface due to the nonlinear joint characteristics. Since the relative displacement is too small compared with the launching system response, its influence on the system response is insignificant. However, the nonlinearity brought by the clamp band joint may affect the stability of the launching system, which requires further investigation.

Acknowledgements

The authors are grateful to the National Natural Science Foundation of China (NNSFC) under Grant Nos. 11072123 and 11102096 for financial support in this study.

References

[1] Pan Z W, Xing Y F, Zhu L W, Dong K and Sun M 2010 Liquid propellant analogy technique in dynamic modeling of launch vehicle *Sci. China Technol. Sc.* **53** 2102-10

[2] Trikha M, Gopalakrishnan S and Mahapatra D R 2011 Analysis of dynamic stability of space launch vehicles under aerodynamic forces using CFD derived data *Int. J. Comput. Methods Eng. Sci Mech.* **12** 213-24

[3] Du W, Wie B and Whorton M 2008 Dynamic modeling and flight control simulation of a large flexible launch vehicle *AIAA Guidance, Navigation and Control Conf. and Exhibit* (*Honolulu, Hawaii, 18-21 August 2008*) 1-25

[4] Esbati A H and Irani S 2009 Structural dynamic stability of three stages launch vehicle as a free-free Timoshenko beam subjected to follower force *Proc. of 4th Int. Conf. on Recent Advances Space Technologies* (*Istanbul, Turkey, 11-13 June 2009*) 150-5

[5] Qin Z Y, Yan S Z and Chu F L 2012 Influence of clamp band joint on dynamic behavior of launching system in ascent flight *P. I. Mech. Eng. G: J. Aer.* doi: 10.1177/0954410012468070

[6] Kim T C, Rook T E and Singh R 2005 Super- and sub-harmonic response calculations for a torsional system with clearance nonlinearity using the harmonic balance method *J. Sound Vib.* **281** 965-93

[7] Von Groll G and Ewins D J 2001 The harmonic balance method with arc-length continuation in rotor/stator contact problems *J. Sound Vib.* **241** 223-33

[8] Guskov M, Sinou J-J and Thouverez F 2008 Multi-dimensional harmonic balance applied to rotor *Mech. Res. Commun.* **35** 537-45

[9] Jaumouillé V, Sinou J-J and Petitjean B 2010 An adaptive harmonic balance method for predicting the nonlinear dynamic responses of mechanical systems - Application to bolted structures *J. Sound Vib.* 329 4048-67

[10] Cameron T M and Griffin J H 1989 An alternating frequency/time domain method for calculating the steady-state response of nonlinear dynamic systems *ASME J. Appl. Mech.* **56** 149-54

[11] Chen G H 1999 Introduction of Long March launch vehicles: LM-3 (Part V) *Aerospace China* **7** 13-4, 16

Motion control and its ground-based experiment of a tethered subsatellite with a controllable rigid arm

Dongping Jin [1], **Hui Chen** [1,2] **and Ti Chen** [1]

[1] State Key Laboratory of Mechanics and Control of Mechanical Structures, Nanjing University of Aeronautics and Astronautics, Nanjing, China.
[2] 94936 PLA Troops, China.

Email: jindp@nuaa.edu.cn

Abstract. The paper presents an experimental research on motion control of a ground-based tethered subsatellite with a controllable rigid arm between the subsatellite and the tether. The forces acting on the tethered subsatellite are first obtained through the dynamic similarity between the subsatellite and a tethered satellite orbiting the Earth. And then, an experimental system for simulating the tethered subsatellite is designed and elaborated in detail, including the airflow embranchment, the measuring and control subsystems, the power and communication subsystems, et al. Furthermore, the operating principle of the ground-based experimental system is illustrated, and the applicability of which is validated by a test of an on-orbit nonlinear optimal control problem.

1. Introduction

The Tethered Satellite System (TSS), which is deployed from the other flying vehicle in space via a tether, is a new technology for artificial gravity creation, LEO to GEO tethers transportation, scientific and technological applications, electro-dynamic study, and so on. As an extension of the application, TSS with an extra offset mechanism on board/subsatellite has been widely used for its highly controllability and repeatability.

Modi and Misra had committed themselves to investigate the dynamics and control of the TSS with an offset from the end of 20th century. They first introduced a three-dimensional offset on a plate-type space station to study the deployment/retrieval dynamics of the TSS in 1987 (Lakshmanan et al., 1987). The controllability of the linearized equations was analyzed numerically, and the control was achieved through a simple velocity feedback using thrusters and momentum wheels. Later, they (Lakshmanan et al., 1989) applied the tension control strategies to the same TSS and found that the proposed scheme damped a given disturbance in a shortest time during station-keeping process and, that the offset control is the most efficient one in terms of energy consumption. Afterwards, Modi and Pidgeon (Modi and Pidgeon, 1994) used the linear quadratic regulator in conjunction with the platform, on which a momentum gyros mounted, to alleviate the longitudinal oscillations of tether during retrieval phase, and designed a ground-based experimental facility to demonstrate the validity of the mathematical model with offset control strategy (Modi et al., 1990; Pradhan et al., 1996).

Nohmi et al. (Nohmi et al., 1998; Nohmi, 2004; Takehara et al., 2008; Nohmi et al., 2007; Nohmi et al., 2010) employed a mechanism with an arm on subsatellite to create tether tension torque. Microgravity experiments showed that the rotation motion can be suppressed by the tether tension torque. Kumar (Kumar and Kumar, 2001) investigated offsets mounted on both the platforms instead of on one end and developed an open-loop strategy and showed the feasibility of those offsets by numerical simulations. In specific missions, however, the highly precision attitude of a satellite is required, such as like NASA's SPECS (Submillimeter Probe of the Evolution of Cosmic Structure) mission, stellar interferometry. While the classical control algorithms were available for controlling the motions of the end-body of tethered satellite system during attitude maneuver, yet it took long time to get the desired states, requiring a high level of energy efforts. For example, the maximum moment for controlling the attitude motion of a tethered subsatellite would arrive at about 0.24Nm, which largely exceeded the usual moment in use of momentum wheel (Wang et al., 2010).

In order to achieve the highly control accuracy with less energy consumption, a controllable rigid arm between the subsatellite and the tether is proposed as a control strategy in this paper. The Gauss pseudospectral method and nonlinear programming are employed in motion control. The dynamic equivalence relative to a tethered satellite orbiting the Earth are detailed. The numerical and experimental results show the effectiveness of the proposed method and control scheme, according to the use of less energy.

2. Dynamic equivalence

2.1. Derivation of nonlinear equations on orbit

Consider the in-plane dynamics of a tethered subsatellite deployed/retrieved from a mother satellite. The subsatellite is envisioned to be a disk of radius r and connected to the tether by a massless controllable arm of length λ at points A and B shown in Figure 1. The coordinate frames of reference are defined as follows: the earth-centered Cartesian inertial frame OXY with unit vectors $\boldsymbol{e}_1$ and $\boldsymbol{e}_2$, and the orbital coordinate frame oxy with unit vectors $\boldsymbol{b}_1$ and $\boldsymbol{b}_2$ placed on the mass center of mother satellite. As shown in the Figure 1, the x-axis is aligned with the local vertical position, and the y-axis is tangential to the orbit. Suppose that the mother satellite moves in an elliptical orbit, which is characterized by the following parameters, i.e., orbital eccentricity e, apogee radius r_a, perigee radius r_p, longitude ascending node $\overline{\Omega}$, orbital inclination angle δ, argument of perigee ϖ, and true anomaly ν. Suppose that the current deployed tether length is l, while the deviation angle of the tether is θ_1, the incremental angle of the controllable arm is θ_2, and the attitude of the subsatellite is φ. Let the tension force of the tether and the torque exerted on the subsatellite is T and M, respectively.

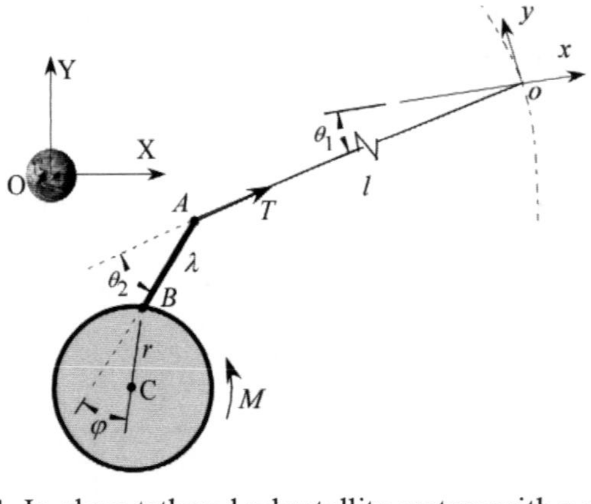

Figure 1. In-plane tethered subsatellite system with a rigid arm

The assumptions employed in the derivation of nonlinear equations are as follows: (1) Solar radiation, air drag force or RF interference from satellite thrusters are negligible. (2) The massless tether is assumed to be a straight, inextensible string. (3) The mass of the mother satellite is assumed to be much larger than the mass of subsatellite m, so that the orbit dynamics of which cannot be influenced by the subsatellite and the tether.

According to Figure 1, it is easy to write out the position vector of the subsatellite in the orbital frame

$$\mathbf{r} = x\boldsymbol{b}_1 + y\boldsymbol{b}_2 = -[l\cos\theta_1 + \lambda\cos\theta + r\cos\phi]\boldsymbol{b}_1 - [l\sin\theta_1 + \lambda\sin\theta + r\sin\phi]\boldsymbol{b}_2 \quad (1)$$

where $\theta=\theta_1+\theta_2$ and $\phi = \theta + \varphi$. The acceleration of the subsatellite is

$$\ddot{\mathbf{r}} = \ddot{\boldsymbol{R}} + (\ddot{x} - \ddot{v}y - 2\dot{v}\dot{y} - \dot{v}^2 x)\boldsymbol{e}_1 + (\ddot{y} + \ddot{v}x + 2\dot{v}\dot{x} - \dot{v}^2 y)\boldsymbol{e}_2 \quad (2)$$

and

$$\ddot{\boldsymbol{R}} = \ddot{R}_x\boldsymbol{e}_1 + \ddot{R}_y\boldsymbol{e}_2 = r_a(1-e^2)\{[(e\cos v - 2e^2\sin^2 v + 1)\dot{v}^2\kappa^{-1} - \ddot{v}e\sin v]\boldsymbol{e}_1 + (2e\dot{v}^2\kappa^{-1}\sin v + \ddot{v})\boldsymbol{e}_2\} \quad (3)$$

where $\dot{v} = [\mu r_a^{-3}(1-e^2)^3]^{1/2}\kappa^2$, $\ddot{v} = -2\mu e\kappa^3 r_a^{-3}(1-e^2)^{-3}\sin v$, $\kappa = 1 + e\cos v$, and μ is the gravitational parameter of Earth. The overdot represents the derivative with respect to time. The gravity force acting on the subsatellite can be described as

$$\boldsymbol{F}_g = F_{gx}\boldsymbol{e}_1 + F_{gy}\boldsymbol{e}_2 \approx m\dot{v}^2[(2x - r_a)\boldsymbol{e}_1 - y\boldsymbol{e}_2] \quad (4)$$

Applying Newton's law to the subsatellite, one obtains

$$m(\ddot{x} - \ddot{v}y - 2\dot{v}\dot{y} - \dot{v}^2 x - \dot{v}^2 r_a) = F_{gx} + T\cos\theta_1 \quad (5a)$$

$$m(\ddot{y} + \ddot{v}x + 2\dot{v}\dot{x} - \dot{v}^2 y) = F_{gy} + T\sin\theta_1 \quad (5b)$$

$$J(\ddot{v} + \ddot{\phi}) = M - T[\lambda\sin\theta_2 + r\sin(\theta_2 + \varphi)] \quad (5c)$$

where J is the moment of inertia of subsatellite.

2.2. Ground-based system

To reveal the dynamics of a tethered satellite system by physical simulation, it needs to construct a ground experiment system that is able to modeling the space environment of the mechanics of the tethered satellite system, according to dynamic similarity. Figure 2 is a schematic diagram of the ground experiment system. The coordinate *oxy* is settled as the same as that of on-orbit tethered satellite system. A tether controller having the measuring capacity of the length and tension of tether is mounted on the origin of the coordinate *oxy* for reeling in/out the tether. The subsatellite, called simulator in what follows, can float on the marble flat by three air-bearings so as to simulate the micro-gravity environment of subsatellite in the direction of deployment or retrieval. The position and attitude of the simulator can be measured by the dynamic measuring system, and controlled by four thrusters and a reaction wheel. The controlling forces are resultant forces F_t and F_r and force moment M_0 as shown in Figure 2. The deflection angle φ of the controllable arm can be controlled by a DC motor.

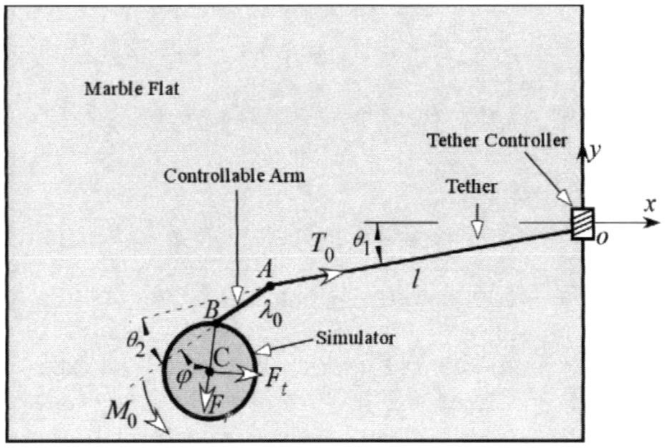

Figure 2. The schematic diagram of ground experiment system

The equations of motion of the simulator is

$$m_0\ddot{x} = F_t \sin\phi - F_r \cos\phi + T_0 \cos\theta_1 \tag{6a}$$
$$m_0\ddot{y} = -F_t \cos\phi - F_r \sin\phi + T_0 \sin\theta_1 \tag{6b}$$
$$J_0\ddot{\phi} = M_0 - T_0[\lambda_0 \sin\theta_2 + r_0 \sin(\theta_2 + \varphi)] \tag{6c}$$

where m_0 and J_0 represent the mass and moment of inertia of simulator, respectively. r_0 is the radius of simulator and λ_0 is the length of controllable arm.

2.3. Equivalence Principle

Comparing Equations (5) with (6), one can find that if let

$$F_r = m_0(\ddot{v}x + 2\dot{v}\dot{x} - \dot{v}^2 y - F_{gy}m^{-1})\sin\phi - m_0(\ddot{v}y + 2\dot{v}\dot{y} + \dot{v}^2 x + \dot{v}^2 r_a + F_{gx}m^{-1})\cos\phi \tag{7a}$$
$$F_t = m_0(\ddot{v}y + 2\dot{v}\dot{y} + \dot{v}^2 x + \dot{v}^2 r_a + F_{gx}m^{-1})\sin\phi + m_0(\ddot{v}x + 2\dot{v}\dot{x} - \dot{v}^2 y - F_{gy}m^{-1})\cos\phi \tag{7b}$$
$$M_0 = \{M - T[\lambda \sin\theta_2 + r\sin(\theta_2 + \varphi)]\}J^{-1}J_0 - J_0\ddot{v} + T_0[\lambda_0 \sin\theta_2 + r_0 \sin(\theta_2 + \varphi)]m^{-1}m_0 \tag{7c}$$
$$T_0 = Tm^{-1}m_0 \tag{7d}$$

there is no difference between the two dynamical systems. The dimensionless equations of motion (6) can be rewritten as

$$l\theta_1'' = m^{\dagger}(Q_x \sin\theta_1 - Q_y \cos\theta_1 - 2\lambda_1 Q_\varphi \xi_2^{-2}) + l\lambda_3 - (2l' + l\ddot{v}\dot{v}^{-2})\theta_1' \tag{8a}$$
$$l\theta_2'' = m^{\dagger}[-Q_x \sin\theta_1 + Q_y \cos\theta_1 + 2(\lambda_1 + 1)Q_\varphi \xi_2^{-2}] - l\lambda_3 - l(\dot{v}^2\varphi'' + \ddot{v}\varphi' - \ddot{v}\theta_2')\dot{v}^{-2} + 2l'\theta_1' \tag{8b}$$
$$l'' = m^{\dagger}(-Q_x \cos\theta_1 - Q_y \sin\theta_1 + 2\lambda_2 Q_\varphi \xi_2^{-2}) + \lambda_4 + l\theta_1'^2 - \ddot{v}l'\dot{v}^{-2} \tag{8c}$$

where

$$\mathrm{d}/\mathrm{d}t = \dot{v}\mathrm{d}/\mathrm{d}v = \dot{v}(\,)' \tag{9a}$$
$$\mathrm{d}^2/\mathrm{d}t^2 = \dot{v}^2\mathrm{d}^2/\mathrm{d}v^2 + \ddot{v}\mathrm{d}/\mathrm{d}v = \dot{v}^2(\,)'' + \ddot{v}(\,)' \tag{9b}$$
$$\xi_1 = \lambda / l \tag{9c}$$
$$\xi_2 = r / l \tag{9d}$$
$$m^{\dagger} = 1/m \tag{9d}$$
$$Q_x = (F_t \sin\phi - F_r \cos\phi + \bar{T}\cos\theta_1)\dot{v}^{-2} \tag{9f}$$
$$Q_y = (-F_t \cos\phi - F_r \sin\phi + \bar{T}\sin\theta_1)\dot{v}^{-2} \tag{9g}$$
$$Q_\varphi = (\bar{M}l^{-1} - \bar{T}[\xi_1 \sin\theta_2 + \xi_2 \sin(\theta_2 + \varphi)])\dot{v}^{-2} \tag{9h}$$
$$\lambda_1 = \xi_1 \cos\theta_2 + \xi_2 \cos(\theta_2 - \varphi) \tag{9i}$$
$$\lambda_2 = \xi_1 \sin\theta_2 + \xi_2 \sin(\theta_2 + \varphi) \tag{9j}$$
$$\lambda_3 = \xi_1(\theta_1' + \theta_2')^2 \sin\theta_2 + \xi_1(\dot{v}^2\varphi'' + \ddot{v}\varphi')\cos\theta_2\dot{v}^{-2} + \xi_2\phi'^2 \sin(\theta_2 + \varphi) \tag{9k}$$
$$\lambda_4 = \xi_1(\theta_1' + \theta_2')^2 \cos\theta_2 - \xi_1(\dot{v}^2\varphi'' + \ddot{v}\varphi')\sin\theta_2/\dot{v}^2 + \xi_2\phi'^2 \cos(\theta_2 + \varphi) \tag{9l}$$

3. Ground-based experimental system

3.1. Experimental setup

Figure 3 shows the experimental platform of the tethered satellite system in State Key Laboratory of Mechanics and Control of Mechanical Structures. The satellite simulator, of diameter 0.2m and mass 12.02kg, can float on the table with the help of three air bearings. It has three degrees of freedom, i.e., two translational motions and one rotation. The tether is made of ultra high molecular weight polyethylene fiber (DYNEEMA), which has extremely long chains with molecular weight numbering in the millions, usually between 2 and 6 million. The air gap between the air bearing and the marble table is less than 5μm and each bearing can sustain a load up to 335N.

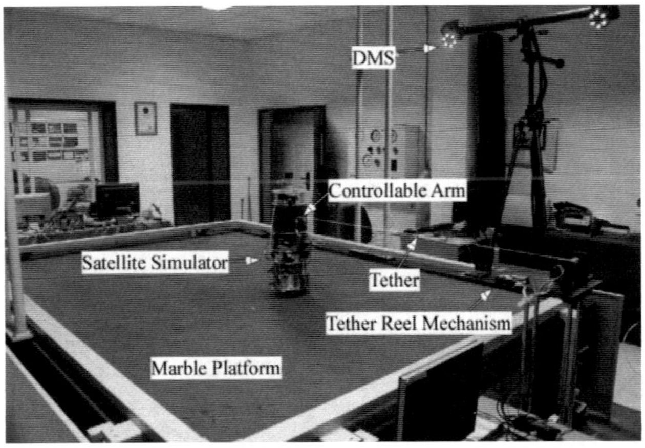

Figure 3. The experimental setup on ground

The ground-based experimental system consists of five subsystems: the pneumatic subsystem, the measuring and control subsystems, the power and communication subsystems. The components of the pneumatic subsystem includes one carbon dioxide tank, one regulator, three air bearings, four pressure-type solenoid valves, four pressure sensors, and four Rafael benitez's nozzles. These parts constitute two airflow embranchments on the simulator, one is for floating and other for control. The airtank is of working capacity of approximately 1 hour for floating and 20 minutes for controlling.

The measuring subsystem can acquire the pressure in front of the valves, the location and the rotation rate of the satellite simulator. A 3D-dynamic-measuring-system (DMS) captures the frame of the ground-based experimental system at a high speed to recognize the markers located on points o, A, B, and C, respectively.

There are four groups of actuators to control the position and attitude of the simulator. The first is the tether reel in/out mechanism, fixed on one side of the marble platform, to control the tether tension or the reel rate. Four solenoid valves, in two orthogonal directions, are mounted on the satellite simulator for position control. The attitude motion is controlled by the reaction wheel placed on the bottom of the simulator. A controllable rigid arm is selected as the offset mechanism for attitude tuning.

Two rechargeable lithium batteries with DC/DC converter are chosen as power supply. They can keep the satellite simulator working for no less than 13 hours. The output voltage and the capacity of one battery are 22.2V and 2600mAh, respectively. The DC/DC converter converts the output voltage to +12V, ±15V, and 5V.

The main function of the communication subsystem is to read out the location information at assigned port generated by the DMS first. Three Digi international XBee OEM RF modules are the embedded solutions for providing wireless end-point connectivity to DMS, computer onboard and the reel mechanism, respectively. The data stream is transmitted in an 8 byte float format according to IEEE 802.15.4 networking protocol for fast point-to-multipoint networking and operate at 57.6kbs each.

3.2. Experimental Principle

Figure 4 illustrates the working flow of the ground-based experimental system. The state of the satellite simulator is first measured by DMS and exchanged in the Intel format. Floating point values are transferred in an 8 byte double format according to IEEE 754. The control commands determined by Equations (7) are sent to the tether reel mechanism and the industrial computer on the satellite simulator synchronously by the wireless modules as well. The satellite simulator floating on the marble platform is simultaneously controlled by the reel motor, the solenoid valves, the reaction wheel, and the offset mechanism. Once the computer gets the signals from the receiver, it sends them to a

Digital Signal Processor (DSP) and two positioning control unit of the motor. The DSP converts the digital signals sent by the computer through serial communications interface (SCI) to analog signals as the inputs of the solenoid valves. The reaction wheel can provide a linear moment to control the attitude motion of the satellite simulator.

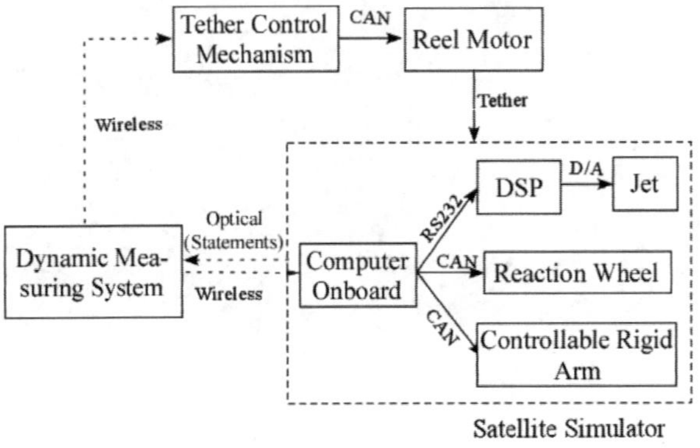

Figure 4. The working flowchart of controlled simulator

3.3. Numerical and experimental results

In this paper, the nonlinear optimal control problem is discretized to a large-scale optimization problem by using the Gauss Pseudospectral method (Benson et al., 2006) , and then the method of nonlinear programming is used to obtain numerical solutions.

Some constraints are imposed on the terminal states, and the cost functional is selected to be

$$J = (u_1^2 + u_2^2 + u_3^2)\mathrm{d}t + \sum_{k=1}^{n}[x_k(t_f) - x_{kf}]^2 \tag{10}$$

where $\mathbf{x} = (\theta_1, \theta_2, \xi, \varphi, \theta_1', \theta_2', \xi', \varphi')$ are the states, n=8 is the number of system states, $\xi = l/L_f$ is the non-dimensional tether length normalized with respect to the fully deployed tether length L_f. The integral term in Equation (10) represents the energy cost during the optimal and the second specification requires that the final states meet some specified endpoint conditions. Solutions are obtained in Matlab 7.3 using GPOPS as the optimization software. The final time is selected as t_f=5 for 30 nodes. Different eccentricities of the orbit are chosen to verify the optimal control method. Some physical constraints are enforced during the optimization process, i.e., the tension forces of the tether are maintained to be positive and less than its tensile strength. Meanwhile, the tether's length is positive but 1.5 times shorter than the fully deployed tether length. The range of the tether angle is from $-\pi/2$ to $\pi/2$ so that the tether cannot swing to the back of the mother satellite. The force moment supplied by reaction wheel is no more than $\pm$0.05Nm.

For simplicity, take the orbital parameters as same as that in Reference (Williams, 2005), namely $(R, \bar{\Omega}, \delta, \varpi, v, e)$= (6770.882km,0,0.5°,0,0,0.2). The initial and terminal states are set as $(\theta_1, \theta_1', \theta_2, \theta_2', \xi, \xi', \varphi, \varphi')|_{t=t_0}$ =(0,0,0,0,1,0,π,0) and $(\theta_1, \theta_1', \theta_2, \theta_2', \xi, \xi', \varphi, \varphi')|_{t=t_f}$ =(0.042,1.4973,0,0,1.0152,0,π,0). The experimental parameters are listed in Table 1. The profiles of the states and the control inputs are illustrated in Figure 5 and Figure 6. As shown in Figure 6, the tether tension force is revealed to be the primary energy input in the control process, the maximum of which reaches 11.43N. Also note that the moment of momentum wheel changes gently in the whole procedure, and the maximum of the moment, presented at much less than saturation values, is 3.944mNm only. The experimental control inputs according to Equations (7) are showed in Figure 7. The maximum amplitudes of the tangential and normal forces are F_r=−0.1291N and F_t=0.0588N, respectively. The maximum and minimum values of the tether tension are 0.22403N and 0.02663N. One can see from Figure 5 that the motion of the tether exhibits an sinusoid-like oscillation in the process of optimization and the maximum amplitude of the oscillation is −0.413rad. As a result of the initial errors between the numerical simulation and the experiment, the satellite simulator gradually increases to the designation.

Table 1. Simulation parameters

Symbol	Pendulum	Experimental
Disc mass m (kg)	500	12.02
Tether length L_f (m)	100	1.0689
Disc radius r (m)	1	0.101
Controllable arm length λ (m)	2	0.1
Moment of inertia J (kg m^2)	0.06	0.035725

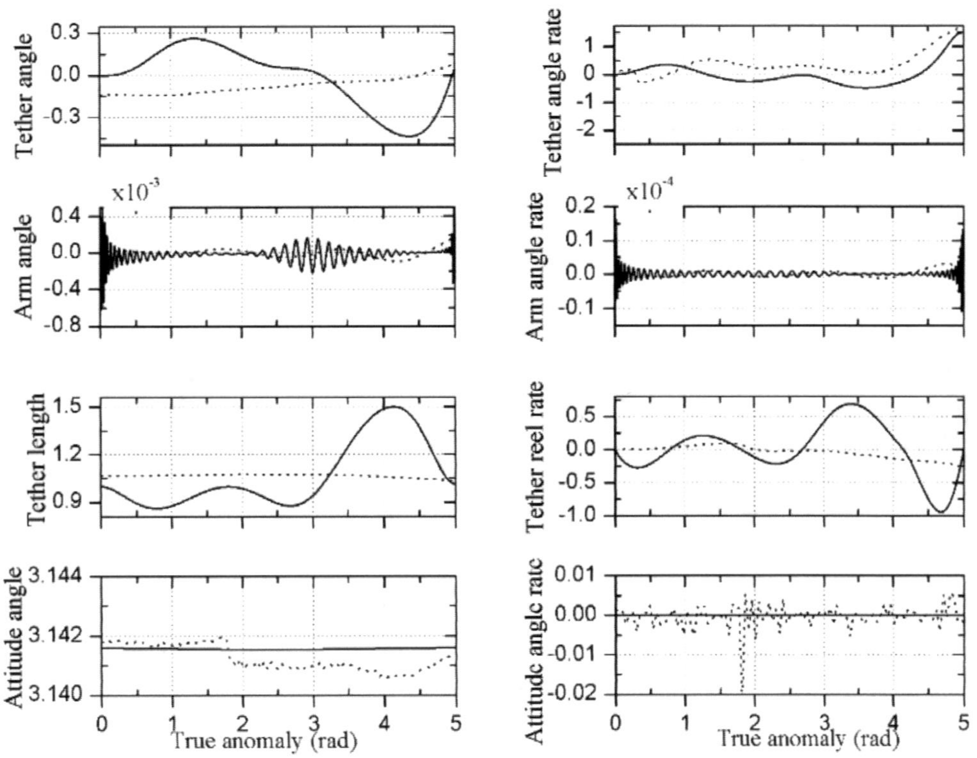

Figure 5. Profiles of the states(—Numerical，…Experimental)

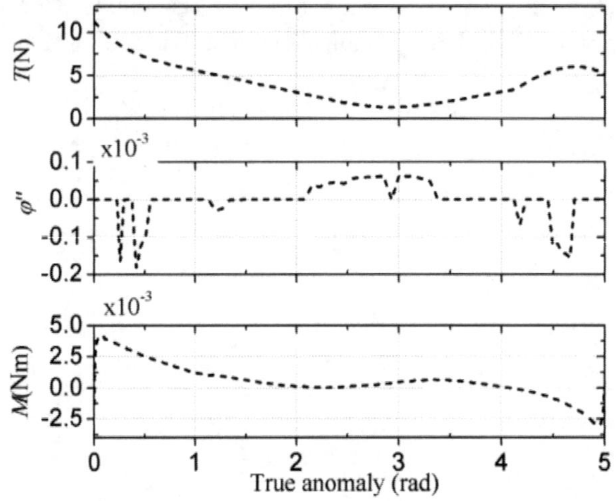

Figure 6. Profiles of the numerical control inputs

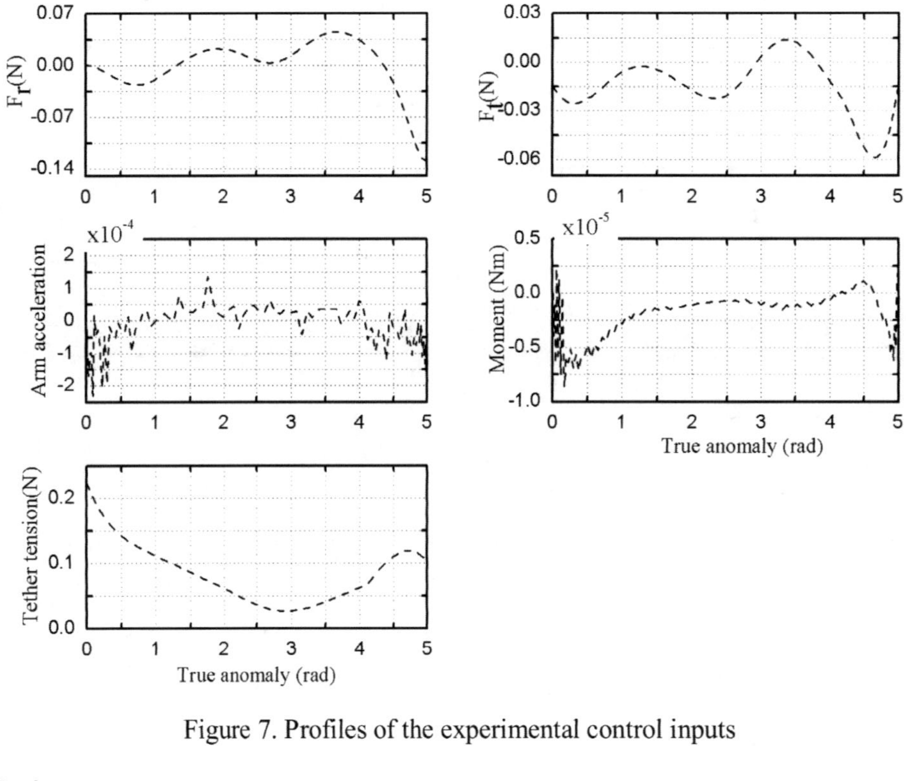

Figure 7. Profiles of the experimental control inputs

4. Conclusions

The possibility of using a controllable rigid arm connected to satellite for motion control of tether satellite system is numerically and experimentally studied. The dynamic equivalent principle and the experimental setup are presented. To show the controllable rigid arm is quite efficient in decreasing the control moment of reaction wheel, the numerical simulations are first made. The controlled satellite simulator can be brought to the destination at a preassigned motion on the marble table.

References

[1] Lakshmanan P K, Modi V J and Misra A K 1987 Dynamics and control of the tethered satellite system in the presence of offsets *Acta Astronautica* 15(12):1053-57.

[2] Lakshmanan P K, Modi V J and Misra A K 1989 Dynamics and control of the tethered satellite system in the presence of offsets *Acta Astronautica* 19(2):145-160.

[3] Modi V J and Pidgeon R P 1994 Dynamics and control of a flexible tethered system with offset *Acta Astronautica* 32(4):255-265.

[4] Modi V J, Lakshmanan P K and Misra A K 1990 Offset control of tethered satellite system: Analysis and experimental verification *Acta Astronautica* 21(5):283-294.

[5] Pradhan S, Modi V J and Misra A K 1996 On the offset control of flexible nonautonomous tethered two-body systems *Acta Astronautica* 38(10):783-801.

[6] Nohmi M, Nenchev D N and Uchiyama M 1998 Momentum control of a tethered space robot through tether tension control *Proceedings of the 1998 IEEE International Conference on Robotics & Automation*, Leuven, Belgium.

[7] Nohmi M 2004 Attitude control of a tethered space robot by link motion under microgravity *Proceedings of the 2004 IEEE Intemational Conference on Control Applications*, Taipei, Taiwan.

[8] Takehara S, Terumichi Y, Nohmi M and Sogabe K 2008 Numerical and experimental approaches on the motion of a tethered system *Journal of System Design and Dynamics* 2(5):1106-17.

[9] Nohmi M, Yamamoto T and Takagi Y 2007 Microgravity experiment for attitude control of a tethered body by arm link motion *Proceedings of the 2007 IEEE International Conference on Mechatronics and Automation*, Harbin, China.

[10] Nohmi M, Tanikawa J and Hosoda T 2010 Simulation analysis of a tethered space robot for space experiment on sounding rocket by JAXA/ISAS *Proceedings of the 2010 IEEE International Conference on Mechatronics and Automation*, Xi'an, China.

[11] Kumar K D and Kumar K 2001 Attitude maneuver of dual tethered satellite platforms through tether offset change *Journal of Spacecraft and Rockets* 38(2):237-242.

[12] Wang X Y, Wen H and Jin D P 2010 Receding horizon control of retrieval of a tethered subsatellite with attitude *Chinese Journal of Theoretical and Applied Mechanics* 42(5):919-925.

[13] Benson D A, Huntington G T, Thorvaldsen T P and Rao A V 2006 Dirrect trajectory optimization and costate estimation via an orthogonal collocation method *Journal of Guidance, Control, and Dynamics* 29(6):1435-40.

[14] Williams P 2005 Spacecraft rendezvous on small relative inclination orbits using tethers *Journal of Spacecraft and Rockets* 42(6):1047-60.

Spatially complex localization in twisted elastic rods constrained to an elliptic cylinder

Jingjing Feng[1,2], Qichang Zhang[1,2] and Wei Wang[1,2]

[1] Department of Mechanics, School of Mechanical Engineering, Tianjin University, Tianjin, China

[2] Tianjin Key Labortory of Nonlinear Dynamics and Chaos Control, Tianjin University, Tianjin, China

E-mail: jjfeng@tju.edu.cn

Abstract. In molecular biology, the elastic rod is wound on the elliptic cylinder can be used as a simplified model of the DNA and protein molecules, in which the elliptic cylinder is used to simulate the heterogeneity of the protein molecules. The problem can be described as a long thin weightless rod constrained, by suitable distributed forces, to lie on an elliptic cylinder while being held by end tension and twisting moment. The Cosserat director theory is used to formulate this problem. More complicated shaped are possible and special attention is given to localized configurations described by homoclinic and heteroclinic orbits of the oscillator. The analytical results of homoclinic and heteroclinic boundary conditions are obtained from Padé approximation. By using the analytical results of homoclinic and heteroclinic solution, both internal force and 3D configuration are discussed in detail. The results of analytical and numerical integration are compared to verify the effectiveness and feasibility of the analytical method.

1. Introduction

In molecular biology, wounded elastic rod in a cylinder can be used as a simplified model of the combination of DNA and protein molecules. This model also has a wide range of applications in other areas, for example, considering drill strings constrained to lies within a cylindrical bore hole [1, 2] in the oil drilling industry. Taking the heterogeneity of the protein molecules into account, we can simplify the model into elastic rod constrained to lie on an elliptic cylinder, where non-circular cross-section of elliptic cylinder is used to simulate protein molecules

The localized buckling of long, thin, initially straight, elastic rods subject to end forces and moments for the case of rods with non-symmetric cross section was studied numerically [3]. The circle of homoclinic orbits present in the symmetric case breaks up and four isolated primary homoclinic orbits were left in breaking the rotational symmetry by considering rods with different bending stiffnesses in two orthogonal directions [4]. The heteroclinic saddle connections are found to play an important role in the post-buckling behavior by defining critical loads at which a straight isotropic rod may coil up into a helix [5]. The anisotropic rod constrained to a cylinder was analyzed by numerical methods [6]. And same results are similar with the isotropic rod constrained to the plane[7].

In the construction of the homoclinic and heteroclinic orbits in nonlinear dynamical systems, numerical simulations are used in most cases. While on contrast the analytical approach. The analytical approaches are rare for constructing the homoclinic and heteroclinic orbits in nonlinear dynamical systems, and numerical simulations are frequently used. The Padé approximants were successfully used for the homoclinic orbits of two-dimensional axisymmetric breathers by Vakakis [8]. The quasi-Padé approximants were used to construct the homoclinic orbits in nonlinear Schrodinger equation system by Mikhlin [9]. The analytic homoclinic orbits, in the non-autonomous Duffing equation and the Van Der Pol-Duffing equation with weakly coupled nonlinear oscillators, were computed in Ref. [10]. The Padé approximation is extended to be used to construct the homoclinic and heteroclinic orbits in the asymmetric systems, which to improve the accuracy of the threshold for the onset of chaos [11-13]. However, the problem of effective analytic approximate of the heteroclinic orbits is difficult and it is not solved up to now.

Attention is also given to localized rod solutions corresponding to homoclinic orbits of the oscillator. But the heteroclinic orbits are special phenomenon which is not presented in the free rod case. They play an important role in the behavior of solutions by defining critical loads at which a straight rod may coil up into a helix. According to the radius of the elliptic cylinder, the heteroclinic orbits occur.

2. Model

The equilibrium equations of the constrained rod are given by[14]

$$\begin{cases} \dot{\vec{n}} = \vec{f}, \\ \dot{\vec{m}} + \dot{\vec{r}} \times \vec{n} = 0, \\ \dot{\vec{r}} = \vec{d}_3, \\ \dot{\vec{d}}_i = \vec{u} \times \vec{d}_i, \end{cases} \tag{1}$$

where $\vec{n}$ and $\vec{f}$ are internal forces and moments along the rod respectively, and $\{\vec{d_1}, \vec{d_2}, \vec{d_3}\}$ is the director frame. The vector $\vec{d_3}$ is taken to be tangent to the rod, where the vector function $\vec{r}$ describes the centerline of the rod relative to some fixed co-ordinate system to be specified later. And the directors $\vec{d_1}$ and $\vec{d_2}$ are chosen along the principal axes of the second moment of area in the normal cross-section of the rod.

The generalized strain vector

$$\vec{u} = u_1\vec{d}_1 + u_2\vec{d}_2 + u_3\vec{d}_3, \tag{2}$$

satisfies the generalized stresses and strains

$$u_1 = \frac{1}{B_1}\vec{m}\cdot\vec{d}_1, \qquad u_2 = \frac{1}{B_2}\vec{m}\cdot\vec{d}_2, \qquad u_3 = \frac{1}{C}\vec{m}\cdot\vec{d}_3, \tag{3}$$

where B_1 and B_2 are the bending stiffnesses about $\vec{d}_1$ and $\vec{d}_2$, respectively, and C is the torsional stiffness.

Considering $\{\vec{e_1}, \vec{e_2}, \vec{e_3}\}$ is a fixed right-handed orthonormal co-ordinate system, we introduce the position vector of the elliptic cylindrical coordinates (R, ψ, z) based on the cylindrical coordinate

$$\vec{r} = R\cos\psi\vec{e}_1 + R\sin\psi\vec{e}_2 + z\vec{e}_3, \tag{4}$$

which meet

$$\begin{cases} R = r\sqrt{\cos^2\psi_0 + \dfrac{b^2}{a^2}\sin^2\psi_0}, \\ \tan\psi = \dfrac{b}{a}\tan\psi_0, \end{cases} \tag{5}$$

where a is the semi-mahor axis of unit elliptical column in $\vec{e}_1$ direction and b is the semi-minor axis in $\vec{e}_2$ direction. And circumferential angle of unit cylinder ψ_0 is introduced. After that, elliptical coordinate system $\{\vec{e_r},\vec{e_\psi},\vec{e_z}\}$ can be written in the following form,

$$\begin{cases} \vec{e_r} = a\cos\psi_0\vec{e_1} + b\sin\psi_0\vec{e_2}, \\ \vec{e_\psi} = -a\sin\psi_0\vec{e_1} + b\cos\psi_0\vec{e_2}, \\ \vec{e_z} = \vec{e_3}, \end{cases} \tag{6}$$

where $\vec{e_r}$ is normal to the cylinder, $\vec{e_\psi}$ is the circumferential direction, and $\vec{e_z}$ is the direction of the axis of the elliptical column. And there are

$$\begin{aligned} \vec{r} &= ra\cos\psi_0\vec{e}_1 + rb\sin\psi_0\vec{e}_2 + z\vec{e}_3 \\ &= r\vec{e}_r + z\vec{e}_z, \end{aligned} \tag{7}$$

which also meets the following changes

$$\begin{cases} \dot{\vec{e}}_r = -\dot{\psi}_0 a\sin\psi_0\vec{e}_1 + \dot{\psi}_0 b\cos\psi_0\vec{e}_2 = \dot{\psi}_0\vec{e}_\psi, \\ \dot{\vec{e}}_\psi = -\dot{\psi}_0 a\cos\psi_0\vec{e}_1 - \dot{\psi}_0 b\sin\psi_0\vec{e}_2 = -\dot{\psi}_0\vec{e}_\psi. \end{cases} \tag{8}$$

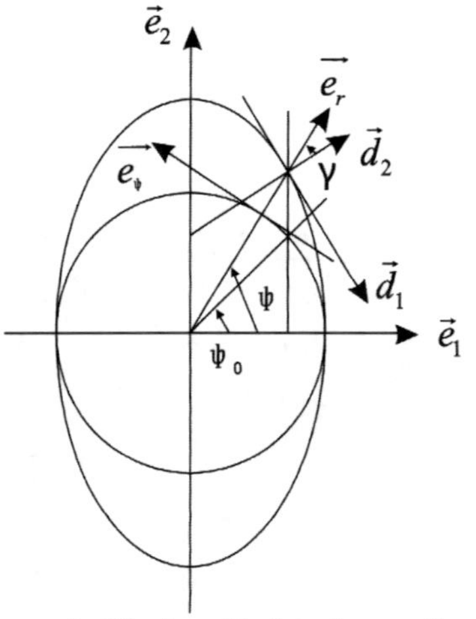

Figure 1 elliptic cylindrical coordinate

Define $\vec{n} = n_r\vec{e}_r + n_\psi\vec{e}_\psi + n_z\vec{e}_z$, $\vec{m} = m_r\vec{e}_r + m_\psi\vec{e}_\psi + m_z\vec{e}_z$, $\vec{d}_i = d_{ir}\vec{e}_r + d_{i\psi}\vec{e}_\psi + d_{iz}\vec{e}_z$ $(i=1,2,3)$. From $\dot{\vec{r}} = \vec{d}_3$ we have

$$\begin{cases} \dot{\psi}_0 = \dfrac{d_{3\psi}}{r}, \\ \dot{r} = d_{3r}, \end{cases} \tag{9}$$

from the components of $\vec{d}_3$, the following equations (10) can be obtained. The external normal reaction force can be written as $\vec{f} = f\vec{e}_r$.

$$\begin{cases} \dot{d}_{3r} = \dfrac{d_{3\psi}^2}{r} + u_2 d_{1r} - u_1 d_{2r}, \\ \dot{d}_{3\psi} = -\dfrac{d_{3r} d_{3\psi}}{r} + u_2 d_{1\psi} - u_1 d_{2\psi}, \\ \dot{d}_{3z} = u_2 d_{1z} - u_1 d_{2z}. \end{cases} \tag{10}$$

Constraint condition is that along the centerline of the elliptical column conserved $\dot{r} = d_{3r} \equiv 0$, namely

$$\frac{d_{3\psi}^2}{r} + u_2 d_{1r} - u_1 d_{2r} \equiv 0. \tag{11}$$

Considering $\dot{\vec{n}} = \vec{f}$, equation is obtained,

$$\begin{cases} \dot{n}_r = f + \dfrac{d_{3\psi} n_\psi}{r}, \\ \dot{n}_\psi = -\dot{\psi}_0 n_r, \\ \dot{n}_z = 0. \end{cases} \tag{12}$$

Introduce a new function h which is defined by

$$\dot{h} = f + \frac{d_{3\psi} n_\psi}{r}, \tag{13}$$

and the transformation is,

$$\begin{cases} \overline{n}_r = n_r - h, \\ \overline{n}_\psi = n_\psi, \\ \overline{n}_z = n_z. \end{cases} \tag{14}$$

That is $\dot{\overline{n}}_r = 0$ and $\dot{\overline{n}}_z = 0$. Considering the side force, $\overline{n}_z = T$ is obtained. And let $\overline{n}_r = 0$, h can be determined in the equation(13). Then moment equation is

$$\dot{\vec{m}} = -\vec{d}_3 \times \overline{n} - h\vec{d}_3 \times \vec{e}_r. \tag{15}$$

The director frame $\{\vec{d}_1, \vec{d}_2, \vec{d}_3\}$ can be expressed in the following form with Euler angle θ, ϕ,

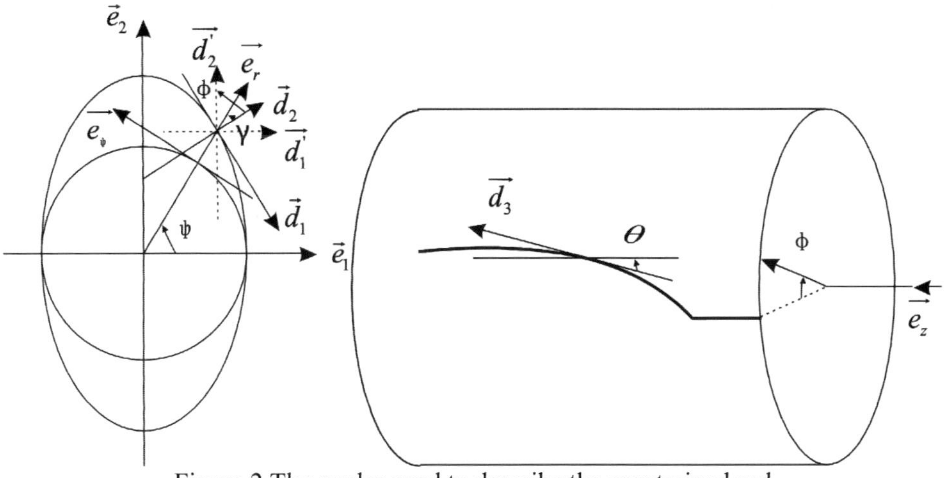

Figure 2 The angles used to describe the constrained rod

$$\begin{cases} \vec{d}_1 = -\sin(\gamma-\phi)\vec{e}_r - \cos(\gamma-\phi)\cos\theta\vec{e}_\psi + \cos(\gamma-\phi)\sin\theta\vec{e}_z, \\ \vec{d}_2 = \cos(\gamma-\phi)\vec{e}_r - \sin(\gamma-\phi)\cos\theta\vec{e}_\psi + \sin(\gamma-\phi)\sin\theta\vec{e}_z, \\ \vec{d}_3 = \sin\theta\vec{e}_\psi + \cos\theta\vec{e}_z, \end{cases} \tag{16}$$

where θ measures deflection angle from centerline of the rod to centerline of elliptic cylinder, and ϕ is internal twist angle of $\vec{d_1}$ turns around $\vec{d_3}$. If rod is not internal twist, then $\phi = 0$. $\vec{d_1}$ is in tangent plane of elliptic cylinder and $\vec{d_2}$ is the corresponding normal line.

The following Eular angle expressions can be had by taking equation (16) into original equation (1),

$$\begin{cases} \dot{\theta} = u_1\sin(\gamma-\phi) - u_2\cos(\gamma-\phi), \\ \dot{\phi} - \dot{\gamma} = u_3 - \dfrac{1}{r}\sin\theta\cos\theta. \end{cases} \tag{17}$$

Introducing the dimensionless variable,

$$\begin{aligned} &x_1 = \frac{n_r}{T}, x_2 = \frac{n_\psi}{T}, x_3 = \frac{n_z}{T}, x_4 = \frac{m_r}{M}, x_5 = \frac{m_\psi}{M}, x_6 = \frac{m_z}{M}, \\ &t = \frac{M}{B_2}s, \tilde{h} = \frac{h}{T}, \tilde{f} = \frac{B_2 f}{MT} \end{aligned} \tag{18}$$

where T and M are the applied loads. According to the derivation process in front, $x_1 = \tilde{h}, x_3 = 1$ can be gained. Finally, the system (1) can be rewritten as the following dimensionless equation

$$\begin{cases} \dot{x}_2 = -\dfrac{1}{\tilde{r}}\tilde{h}\sin\theta, \\ \dot{x}_4 = -\dfrac{1}{m^2}\sin\theta + \dfrac{1}{m^2}x_2\cos\theta + \dfrac{1}{\tilde{r}}x_5\sin\theta, \\ \dot{x}_5 = -\dfrac{\tilde{h}}{m^2}\cos\theta - \dfrac{1}{\tilde{r}}x_4\sin\theta, \\ x_6 = \dfrac{\tilde{h}}{m^2}\sin\theta, \\ \dot{\theta} = -x_4 - \rho(x_4\sin^2(\gamma-\phi) + x_5\sin(\gamma-\phi)\cos(\gamma-\phi)\cos\theta - \\ \qquad x_6\sin(\gamma-\phi)\cos(\gamma-\phi)\sin\theta), \\ \dot{\phi} - \dot{\gamma} = (1+\nu)(x_5\sin\theta + x_6\cos\theta) - \dfrac{1}{\tilde{r}}\sin\theta\cos\theta, \end{cases} \tag{19}$$

where

$$m = \frac{M}{\sqrt{B_2 T}}, \rho = \frac{B_2}{B_1} - 1, \nu = \frac{B_2}{C} - 1, \tilde{r} = \frac{M}{B_2}r \tag{20}$$

are all dimensionless parameters.

Boundary constraint equation (11) may also be written as

$$\begin{aligned} &x_5\cos\theta - x_6\sin\theta + \rho(x_4\sin(\gamma-\phi)\cos(\gamma-\phi) + \\ &x_5\cos^2(\gamma-\phi)\cos\theta - x_6\cos^2(\gamma-\phi)\sin\theta) + \frac{1}{\tilde{r}}\sin^2\theta = 0. \end{aligned} \tag{21}$$

Here, the solutions of equation (19) can be traced, which satisfies the condition (21). And from the third equation of (1) we can obtain,

$$\begin{cases} \dot{\psi}_0 = \dfrac{1}{\tilde{r}}\sin\theta, \\ \dot{z} = \cos\theta. \end{cases} \tag{22}$$

In the coordinate system $\{\vec{e_1},\vec{e_2},\vec{e_3}\}$ there are

$$\begin{cases} x = R\cos\psi = \tilde{r}a\cos\psi_0, \\ y = R\sin\psi = \tilde{r}b\sin\psi_0. \end{cases} \tag{23}$$

Two integral can be obtained from system (19).

$$\mathrm{I}_1 = \tilde{r}x_2 + m^2 x_6 = \text{constant}. \tag{24}$$

This integral equation represents the axial moment equilibrium. The second integral derivated from the "Hamiltonian"

$$\begin{aligned} \mathrm{I}_2 &= \frac{1}{2}(1+\rho)[-x_4\sin(\gamma-\phi) - x_5\cos(\gamma-\phi)\cos\theta + x_6\cos(\gamma-\phi)\sin\theta]^2 \\ &+\frac{1}{2}[x_4\cos(\gamma-\phi) - x_5\sin(\gamma-\phi)\cos\theta + x_6\sin(\gamma-\phi)\sin\theta]^2 \\ &+\frac{1}{2}(1+\nu)(x_5\sin\theta + x_6\cos\theta)^2 + \frac{\cos\theta}{m^2} + \frac{x_2\sin\theta}{m^2} = \text{constant}. \end{aligned} \tag{25}$$

In circular cross section, the third integral is provided by conservative torque u_3,

$$\mathrm{I}_3 = x_5\sin\theta + x_6\cos\theta = \text{constant}. \tag{26}$$

3. Analysis on the circular cross section of rod

System (19) is integrable which is made from the three integral in circular cross section mentioned above, and after simplification the following differential equations can be obtained,

$$\frac{1}{2}\dot{\theta}^2 + V(\theta) = H, \tag{27}$$

where

$$V(\theta) = \frac{\cos\theta}{m^2} + \frac{K_1\sin\theta}{\tilde{r}m^2} - \frac{K_3\sin\theta\cos\theta}{\tilde{r}} + \frac{\cos^2\theta}{\tilde{r}^2} - \frac{\cos^4\theta}{2\tilde{r}^2}, \tag{28}$$

$$H = K_2 - \frac{1}{2}K_3^2(1+\nu) + \frac{1}{2\tilde{r}^2}. \tag{29}$$

Boundary conditions are K_1, K_2 and K_3, which is the value of first integral I_1, I_2 and I_3 respectively. In arbitrary parameter values condition, the system (19) has trivial solution,

$$x_2 = 0, x_4 = 0, x_5 = 0, x_6 = 1, \theta = 0, \dot{\phi} = 1+\nu, \tag{30}$$

meanwhile $\tilde{h} = 0$. At the same time, according to the boundary conditions, system (19) meet the symmetry

$$\begin{aligned} &R:(x_2,x_4,x_5,x_6,\sin\theta,\sin 2(\phi-\gamma)) \to (x_2,-x_4,x_5,x_6,\sin\theta,-\sin 2(\phi-\gamma)), \\ &t \to -t \quad (h \to -h). \end{aligned} \tag{31}$$

Considering the homoclinic boundary conditions, make the values of K_1 and K_3 into

$$K_1 = m^2, K_3 = 1. \tag{32}$$

And K_2 is a free parameter which is used to adjust the energy H of energy balance oscillator. We also have

$$\tilde{h} = \frac{m^2 x_4}{\tilde{r}}(\tilde{r}x_5\sin\theta + \tilde{r}x_6\cos\theta - 3\sin\theta\cos\theta), \tag{33}$$

and $\tilde{f}(\theta)$,

$$\begin{aligned}\tilde f(\theta)=\frac{1}{\tilde r^3}[&(5\tilde r\sin\theta-15\tilde r\sin\theta\cos^2\theta)K_1-(6m^2\tilde r^2-12m^2\tilde r^2\cos^2\theta)K_2\\&-(\tilde r^3\sin\theta+10m^2\tilde r\sin\theta\cos\theta-20m^2\tilde r\sin\theta\cos^3\theta)K_3\\&+(4m^2\tilde r^2+3m^2\tilde r^2\nu-8m^2\tilde r^2\cos^2\theta-6m^2\tilde r^2\nu\cos^2\theta)K_3^2\\&+\tilde r^2\cos\theta K_1K_3-2m^2+9\tilde r^2\cos\theta+16m^2\cos^2\theta\\&-15\tilde r^2\cos^3\theta-26m^2\cos^4\theta+12m^2\cos^6\theta.\end{aligned}\tag{34}$$

Equation (27) contains rich physical phenomena. Consider the initial conditions of $\theta(0)=0,\ \dot\theta(0)=0$, through numerical analysis, we can see that equation has two critical points respectively, which will produce the phenomenon of heteroclinic orbits, when $0\le\tilde r\le 0.9648$ and $1.9093\le\tilde r\le 2$. Taking $\tilde r=0.8$ for example, the system (27) has two critical load $m_{c1}=1.1177$ and $m_{c1}=1.9499$. The second case is analyzed in this paper and its potential energy curves and phase diagram are shown in Figure 3(a). Most of the existing literature only simulated one homoclinic orbit of system (27) using numerical method and do not analyze the phenomenon of heteroclinic orbits in detail. In the next section, we will apply the analytical method that calculated analytical heteroclinic orbits in two typical critical stats. Heteroclinic solutions and its corresponding 3D configuration and internal force diagrams are obtained respectively.

Let

$$p=\cos\theta.\tag{35}$$

When $0<\theta<\pi$, the original equation (27) can be simplified as follows,

$$\begin{aligned}&\frac{1}{2}\dot p^2+\frac{p}{m^2}+(\frac{3}{2\tilde r^2}+K_2-\frac{1}{2}K_3^2(1+\nu))p^2-\frac{1}{m^2}p^3\\&-\frac{3}{2\tilde r^2}p^4+\frac{1}{2\tilde r^2}p^6-(K_2-\frac{1}{2}K_3^2(1+\nu)+\frac{1}{2\tilde r^2})\\&+(\frac{K_1}{\tilde r m^2}-\frac{K_3}{\tilde r}p-\frac{K_1}{\tilde r m^2}p^2+\frac{K_3}{\tilde r}p^3)\sqrt{1-p^2}=0.\end{aligned}\tag{36}$$

Differentiating both ends of the equation, we can get a second-order differential equation

$$\begin{aligned}&\ddot p+\frac{1}{m^2}+(\frac{2}{m^2}+\frac{3}{\tilde r^2})p-\frac{3}{m^2}p^2-\frac{6}{\tilde r^2}p^3+\frac{3}{\tilde r^2}p^5\\&-(\frac{1}{\tilde r}+\frac{3}{\tilde r}p-\frac{5}{\tilde r}p^2-\frac{3}{\tilde r}p^3+\frac{4}{\tilde r}p^4)\frac{1}{\sqrt{1-p^2}}=0.\end{aligned}\tag{37}$$

When $-\pi<\theta<0$, $\sin\theta=-\sqrt{1-p}$. Simplified equation also can be obtained by using similar solution method, so the detailed solving procedure is not explained again.

4. Padé approximation

Extended Padé approximation method has been successfully used to solve the heteroclinic orbits of asymmetric system. The extended Padé approximation is implemented here. Set the series solution of equation (37) to

$$x(t)=a_0+a_1s+\cdots=\sum_{n=0}^{\infty}a_ns^n.\tag{38}$$

Initial point values of the track are represented as (a_0,a_1), and a_0,a_1 are not zero at the same time,

$$x(0)=a_0,\qquad \dot x(0)=a_1.\tag{39}$$

Other parameters can be expressed as a function of a_0,a_1,r,m:

$$
\begin{aligned}
a_2 &= \frac{1}{2m^4r^4}(-m^2r^4 - 3m^4r^2a_0 - 2m^2r^4a_0 + 3m^2r^4a_0^2 + 6m^4r^2a_0^3 \\
&\quad + \sqrt{(m^8r^6 + 6m^8r^6a_0 - 30m^8r^6a_0^3 + 15m^8r^6a_0^4 + 24m^8r^6a_0^5 - 16m^8r^6a_0^6)}) \\
&\quad - 3m^4r^2a_0^5, \\
a_3 &= (-3m^2r^2a_1 + 3m^4r^2a_1 - 9m^4a_0a_1 - 12m^2r^2a_0a_1 + 45m^2r^2a_0^2a_1 \\
&\quad - 45m^4r^2a_0^2a_1 + 18r^4a_0^2a_1 + 72m^4a_0^3a_1 + 48m^2r^2a_0^3a_1 + 30m^4r^2a_0^3a_1 \\
&\quad - 18r^4a_0^3a_1 - 105m^2r^2a_0^4a_1 + 60m^4r^2a_0^4a_1 - 162m^4a_0^5a_1 - 36m^2r^2a_0^5a_1 \\
&\quad - 48m^4r^2a_0^5a_1 + 63m^2r^2a_0^6a_1 + 144m^4a_0^7a_1 - 45m^4a_0^9a_1 - 6m^4r^2a_1a_2 \\
&\quad - 4m^2r^4a_1a_2 + 12m^2r^4a_0a_1a_2 + 36m^4r^2a_0^2a_1a_2 - 30m^4r^2a_0^4a_1a_2) \\
&\quad -2r^4a_1 + 2r^4a_0a_1 / (6m^2r^2(r^2 + 3m^2a_0 + 2r^2a_0 - 3r^2a_0^2 - 6m^2a_0^3 \\
&\quad + 3m^2a_0^5 + 2m^2r^2a_2), \\
&\cdots
\end{aligned}
\tag{40}
$$

4.1. *Homoclinic and heteroclinic orbits*

Transformed system (37) has homoclinic and heteroclinic orbits at same time, when $r = 0.8, m_{c1} = 1.94988$. Potential energy curve and phase diagram are shown in Figure 3(b), wherein solid lines indicate the graphics of system (37) at $0 < \theta < \pi$ and dashed lines at $-\pi < \theta < 0$. The results at $0 < \theta < \pi$ are analyzed here. Phase diagram has one homoclinic orbit and two heteroclinic orbits which are symmetric about the p-axis but not symmetric about the $\dot{p}$-axis. As shown that H_1 and H_2 are two saddle points, and O is center. When $s \to +\infty$, point on homoclinic orbits move along H_1AH_2 and finally tend to reach saddle point H_1, and point on heteroclinic orbits movement along $H_1B_1H_2$ and $H_2B_2H_1$ respectively and finally tend to reach the corresponding saddle point H_1 and H_2; when $s \to -\infty$, along the opposite direction to the other saddle point.

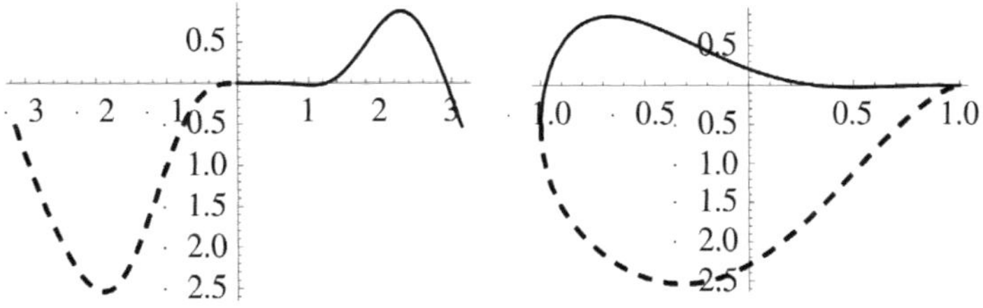

(a) system (27)

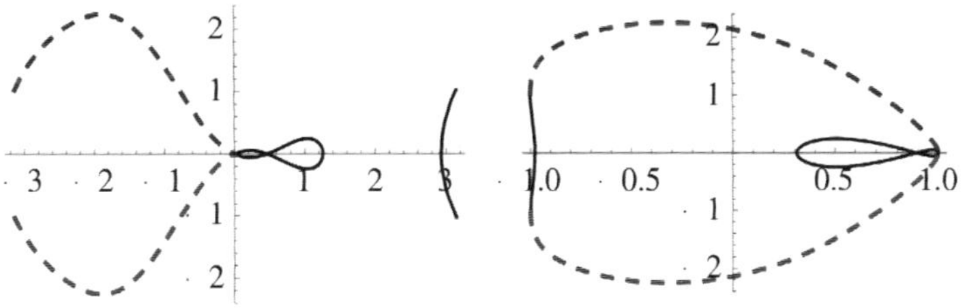

(b) system (37)

Figure 3 Potential energy curve and phase diagram

Take an initial vales are maximum point at p-axis, which meet its tangent perpendicular to the p-axis, namely

$$a_2 = 0. \tag{41}$$

Form of homoclinic solution is

$$PA = \frac{(\alpha_0 + \alpha_1 t + \alpha_2 t^2 + \alpha_3 t^3 + \alpha_4 t^4)}{(1 + \beta_1 t + \beta_2 t^2 + \beta_3 t^3 + \beta_4 t^4)}. \tag{42}$$

All the coefficients of PA_n are obtained by comparing the PA_n and Taylor expansion solution, namely

$$\sum_{n=0}^{\infty} a_n t^n \cong PA_n. \tag{43}$$

Then consider the convergence condition:

$$\lim_{x\to\infty} PA \to H. \tag{44}$$

The other parameters can be expressed by the algebraic polynomial as a_0, a_1 either. And used the convergence condition, the initial value can be obtained respectively. We obtain the initial values $a_0 = 0.310214$, and fourth-order approximation homoclinic solution,

$$p = \frac{0.310214 + 0.257087s^2 + 0.0365497s^4}{1 + 0.347095s^2 + 0.0409154s^4}. \tag{45}$$

Take initial vales at the maximum point of $\dot{p}$-axis, which also meet its tangent perpendicular to the $\dot{p}$-axis, namely

$$a_3 = 0. \tag{46}$$

Form of heteroclinic solution is

$$QPA = \frac{(\alpha_0 + \alpha_1 e^{\sqrt{\omega}t} + \alpha_2 e^{2\sqrt{\omega}t} + \alpha_3 e^{3\sqrt{\omega}t} + \alpha_4 e^{4\sqrt{\omega}t})}{(1 + e^{\sqrt{\omega}t}\beta_1 + e^{2\sqrt{\omega}t}\beta_2 + e^{3\sqrt{\omega}t}\beta_3 + e^{4\sqrt{\omega}t}\beta_4)}, \tag{47}$$

where ω is etermined as the coefficient to be solved. Calculation process is resembled to precious solution. Convergence condition (44) is considered, and then initial value $a_1 = \pm 0.0145213, \omega = 0.170252$ and heteroclinic solution can be obtained

$$\begin{aligned} p^+ &= \frac{0.8933 + 4.06384e^{0.412616s} + 3.89703e^{0.825232s} + 1.61965e^{1.23785s} + 0.268084e^{1.65046s}}{1 + 4.06384e^{0.412616s} + 3.89703e^{0.825232s} + 1.61965e^{1.23785s} + 0.268084e^{1.65046s}}, \\ p^- &= \frac{1 + 6.04159e^{0.412616s} + 14.5366e^{0.825232s} + 15.1588e^{1.23785s} + 3.33217e^{1.65046s}}{1 + 6.04002e^{0.412616s} + 14.6258e^{0.825232s} + 16.5536e^{1.23785s} + 3.73018e^{1.65046s}}. \end{aligned} \tag{48}$$

The calculated position of the saddle point is 0.8933 and 1 which are identical with exact values. Compare present solution of homoclinic and heteroclinic solutions about arc length s and homoclinic and heteroclinic orbits with numerical solution as shown in Figure 4 - 6.

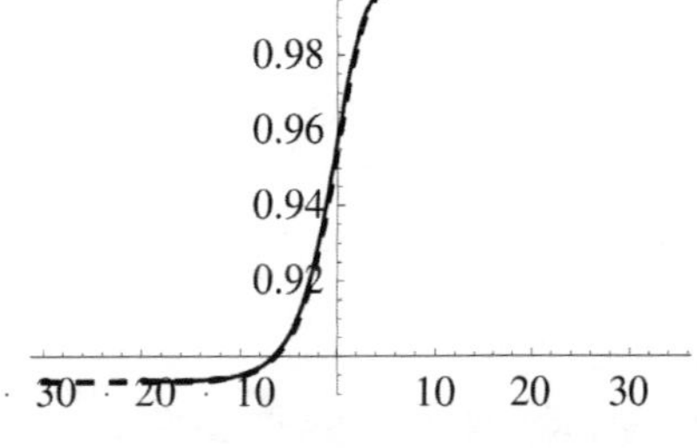

(a) $a_1 > 0$

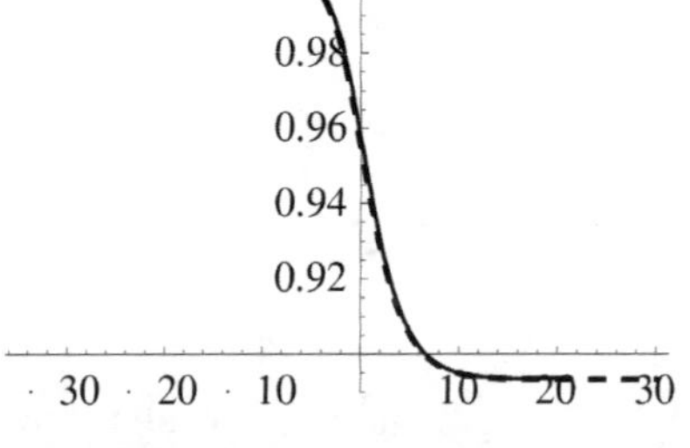

(b) $a_1 < 0$

Figure 4 heteroclinic solution

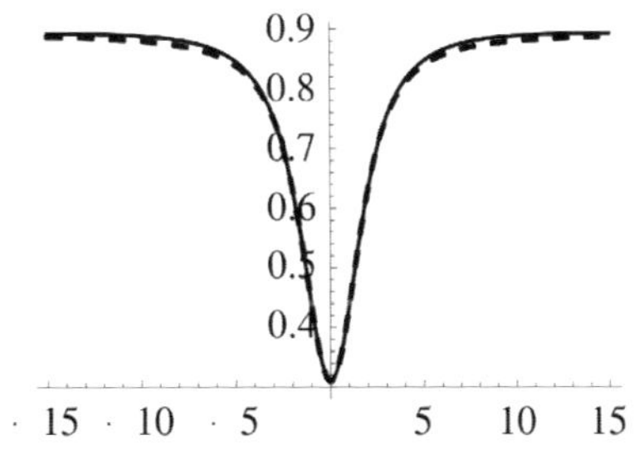

Figure 5 homoclinic solution

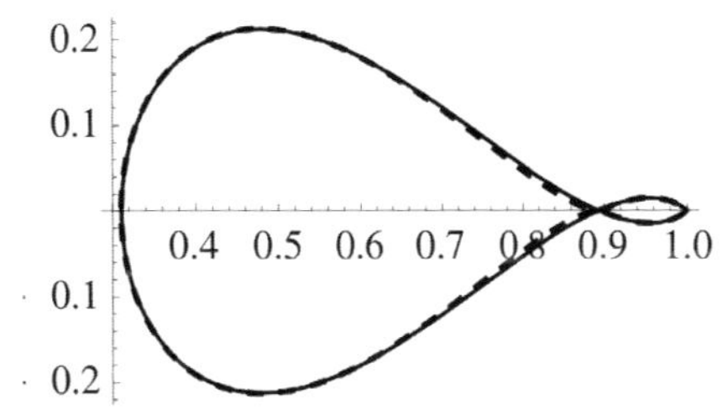

Figure 6 homoclinic and heteroclinic orbits

The rod of internal force and 3D configuration can be obtained respectively. When $s \to \infty$, homoclinic orbit appears, θ is gradually decreased from 1.25538 to 0.466162 and then remains constant. In the corresponding 3D configuration, rotation angle, beginning with the location of the initial point 0.315416, gradually increases to 1.10463, then spiral extends to the right endsand hang with this angle, while the state of rod remain stretching. The axial force of rod has maximum value located at initial point and point to outer cylinder. As $s \to \infty$, axial force gradually decreases to zero. As shown in Figure 7, rotational angle of the rod increases to the ends and spiral extending with constant angle. When heteroclinic orbits appear, taking $a_1 > 0$ for example. When $s \to +\infty$, θ is gradually changed from 0.300549 to 0.466162 and then remains constant; when $s \to -\infty$, θ is gradually changed from 0.300549 to 0 and also remains constant. In the corresponding 3D configuration, rotation angle, beginning with the location of the initial point 1.27025, gradually decreases to 1.10463, while state remains stretching. Rotation angle gradually increase to $\pi/2$, namely changed to horizontally extending. The axial force of rod has maximum value in the location of initial point and point to inner cylinder. As $s \to -\infty$, axial force of rod change to point to outer cylinder, and then decreases to zero; as $s \to +\infty$, axial force also has a related change and finally reach to 0.0181901. As shown in Figure 8, at one end rotational angle of the rod increases and spiral extending with constant angle, at other end it extends horizontally. When $a_1 < 0$, the process is similar to the above.

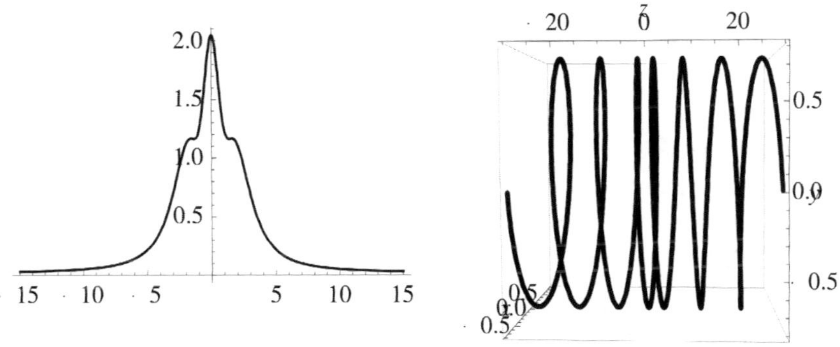

Figure 7 internal force and 3D configuration (homoclinic solution)

Journal of Physics: Conference Series **448** (2013) 012005 doi:10.1088/1742-6596/448/1/012005

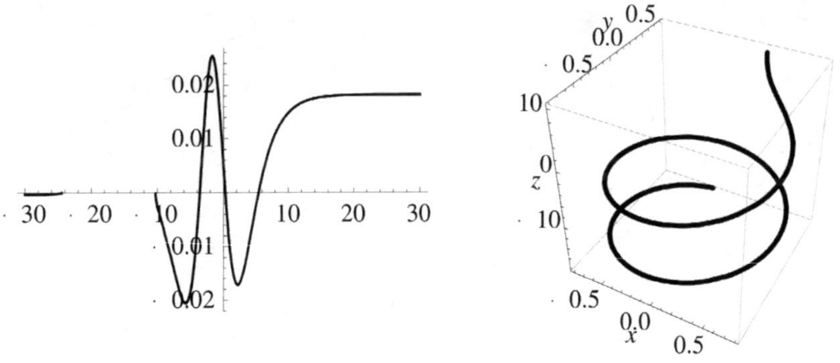

Figure 8 internal force and 3D configuration (heteroclinic solution $a_1 > 0$)

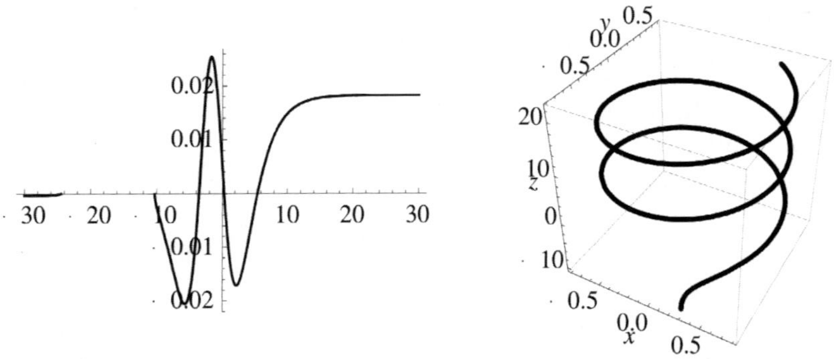

Figure 9 internal force and 3D configuration (heteroclinic solution $a_1 < 0$)

5. Conclusions

This paper presents the general formulation of a twisted rod constrained to lie on an elliptic cylinder. On the statics of twisted rods, much headway has been made by dynamical systems analogy, and homoclinic and heteroclinic orbits are analyzed to describe localized buckling modes of long rods. The strategy presented in this work is generally applied and accepted, so it would be possible to apply the present method to simulate heterogeneity of the protein molecules with more complicated nonlinearity which will be the topic for further research.

By developing the Padé approximation method, the homoclinic and heteroclinic orbits of nonlinear irrational equations can be constructed, which would expand the ranges of the applicable systems, and also can be applied to improve the accuracy of the result. It is feasible to apply this method on the other complex systems; furthermore, by using the combination of high order Melnikov method and improved Padé approximation method, the accuracy of calculation can be improved even further.

Conclusions

Project supported by National Natural Science Foundation of China (Grant No. 11072168, 11102127), Specialized Research Fund for the Doctoral Program of Higher Education of China (Grant No. 20100032120006) and Tianjin Research Program of Application Foundation and Advanced Technology (12JCYBJC12500, 12JCZDJC28000).

References

[1] Bedrossian A N, Abbassian F, Dunayevsky V A and Hope S A. *Analysis of post-buckling behaviour of drillstrings.* in *8th U.K. ABAQUS User Group Conference.* 1993.

[2] Paslay P R and Bogy D B 1964 The stability of a circular rod laterally constrained to be in contact with an inclined circular cylinder *Joural of applied mechanics-transactions of the ASME* **31** 605-10

[3] Heijden G H M, Champneys A R and Thompson J M T 1998 The spatial complexity of localized buckling in rods with noncircular crosssection *SIAM J. Appl. Math.* **59** 198-221

[4] Van Der Heijden G H M and Thompson J M T 1998 Lock-on to tape-like behaviour in the torsional buckling of anisotropic rods *Physica D* **112** 201-24

[5] Van Der Heijden G H M 2001 The static deformation of a twisted elastic rod constrained to lie on a cylinder *Proceedings of the Royal Society A: Mathematical, Physical and Engineering Sciences* **457** 695-715

[6] Van Der Heijden G H M, Champneys A R and Thompson J M T 2002 Spatially complex localisation in twisted elastic rods constrained to a cylinder *Int. J. Solids Struct.* **39** 1863-83

[7] Van Der Heijden G H M, Champneys A R and Thompson J M T 1999 Spatially complex localisation in twisted elastic rods constrained to lie in the plane *Jouranl of the mechanics and physics of solids* **47** 59-79

[8] Emaci E, Vakakis A F, Andrianov I V and Mikhlin Y 1997 Study of two-dimensional axisymmetric breathers using Pade approximants *Nonlinear Dyn.* **13** 327-38

[9] Mikhlin Y V 2000 Analytical Construction of Homoclinic Orbits of Two- and Three-Dimensional Dynamical Systems *J. Sound Vib.* **230** 971-83

[10] Manucharyan G V and Mikhlin Y V 2005 The construction of homo- and heteroclinic orbits in non-linear systems *J. Appl. Math. Mech.* **69** 39-48

[11] Feng J J, Zhang Q C and Wang W 2011 The construction of homoclinic and heteroclinic orbitals in asymmetric strongly nonlinear systems based on the Pade approximant *Chin. Phys. B* **20** 090202

[12] Feng J J, Zhang Q C and Wang W 2012 Chaos of several typical asymmetric systems *Chaos, Solitons Fractals* **45** 950-58

[13] Zhang Q C, Feng J J and Wang W 2011 The construction of homoclinic and heteroclinic orbit in two-dimentsional nonlinear systems based on the quasi-Pade approximation *Chinese Journal of Theoretical and Applied Mechani* **43** 914-21

[14] Antman S S 1995 *Nonlinear problems of elasticity* Springer

Journal of Physics: Conference Series **448** (2013) 012006 doi:10.1088/1742-6596/448/1/012006

Transfer function method for frequency response and damping effect of multilayer PCLD on cylindrical shell

Q Qiu, Z P Fang, H C Wan and L Zheng
The State Key Laboratory of Mechanical Transmission, Chongqing University, Chongqing, China

E-mail: zling@cqu.edu.cn

Abstract. Based on the Donnell assumptions and linear visco-elastic theory, the constitutive equations of the cylindrical shell with multilayer Passive Constrained Layer Damping (PCLD) treatments are described. The motion equations and boundary conditions are derived by Hamilton principle. After trigonometric series expansion and Laplace transform, the state vector is introduced and the dynamic equations in state space are established. The transfer function method is used to solve the state equation. The dynamic performance including the natural frequency, the loss factor and the frequency response of clamped-clamped multi-layer PCLD cylindrical shell is obtained. The results show that multi-layer PCLD cylindrical shell is more effective than the traditional three-layer PCLD cylindrical shell in suppressing vibration and noise if the same amount of material is applied. It demonstrates a potential application of multi-layer PCLD treatments in many critical structures such as cabins of aircrafts, hulls of submarines and bodies of rockets and missiles.

1. Introduction

The Passive Constrained Layer Damping (PCLD) is composed of host shell, visco-elastic material (VEM) layer and constrained layer. In the past decades, it has been an effective way to suppress the vibration and noise in structures. Because the tensile deformation in VEM is limited by the constrained layer, the shear deformation in VEM is increased, more vibration energy is dissipated by means of VEM, vibration and noise in structures is thus reduced significantly. The PCLD is widely used in aviation, aerospace, submarine, vehicle, etc due to simple structure, low cost, high reliability and effectiveness in a wide temperature and frequency range.

In this field, the pioneer work (Kerwin, 1959) derived an expression for an effective, complex, flexural stiffness of the three-layer beams with damping core layer. By extending Kerwin's work, a sixth order partial differential equation of motion governing the transverse displacement of a sandwich beam with arbitrary boundary conditions was developed and a theoretical foundation for PCLD beams was established (Mead et al. 1969). The influence of treatment location on the vibration characteristic of PCLD beams has been investigated (Yeh et al. 2006). By developing a dynamic model of a PCLD laminated rectangular plate, the Galerkin Element Method (GEM) was proposed (Zhang et al. 2000); The GEM is more accurate and effective than traditional Finite Element Method (FEM). A mathematical model for PCLD plate based on the thick plate theory was developed (Chen et al. 2002); Their research suggested that the maximum vibration energy dissipation in PCLD plate can be achieved as the thickness of Constrained Layer (CL) is twice of the thickness of VEM.

The work on the cylindrical shell with PCLD treatment has attracted many researches in the recent decades. The solution mainly included FEM, analytical method and semi-analytical method. FEM has been widely applied due to its flexible and available for arbitrary geometry and boundary conditions. A conventional finite element model for the vibration and damping analysis of a cylindrical shell with a visco-elastic core was developed and the influence of geometric parameters and VEM's shear modulus on suppressing vibration was investigated (Ramesh et al. 1994). The dynamic characteristic of a cylindrical shell with partial PCLD treatment (Wang et al. 2004) was analyzed using a finite element model. A finite element model of the cylindrical shell with fully PCLD treatment was established (Krishna et al. 2007); The influence of the fluid-load on the cylindrical shell in the Bessel function was considered and the nature frequencies and loss factors of a fluid-filled cylindrical shell with PCLD treatment was discussed. A new general motion equation of the PCLD laminated structure by Hamilton principle and Donnell-Mushtari- Vlasov (DMV) assumption was reformulated (Hu et al. 2000). However, different from previous work, their equation only contained three displacement components, therefore its computational efficiency was improved greatly. Although FEM is simple and powerful to analyze the dynamic characteristics of the PCLD cylindrical shell with or without fluid-load, its accuracy is limited to low frequency range due to the adaptation of the low order interpolation function. Meanwhile, FEM is very time consuming when a large or complicated structure is computed. On the other hand, analytical methods for PCLD cylindrical shells are not always available due to its complex and arbitrary boundary conditions. After formulating the motion equation, an analytical solution for the axisymmetric vibration of a circular cylindrical shell with full PCLD treatment was obtained (Pan. 1969). By using the analysis method, the dynamic property of PCLD cylindrical shell has been computed (Cao et al. 2011). Therefore, semi-analytical methods for the PCLD cylindrical shells are further put forward such as assume mode method, modal strain energy method and transfer matrix method. Based on the Donnell thin shell theory and Hamilton principle, the motion equation and boundary conditions of the cylindrical shell with partial PCLD treatment was derived and the state space equation was constructed (Li et al. 2007). The vibration and damping characteristics of the PCLD cylindrical shell with clamped support condition at two ends is investigated by using transfer matrix method. However, the state vector in this approach was formulated by displacement components and their high order derivatives, thus the stress boundary conditions need to be expressed by the displacement components in advance, which was not easy and convenient to apply directly. Recently, a novel control equation of the PCLD cylindrical shell has been formulated and a transfer matrix method combined with highly precise extended homogeneous capacity precision integration technology has been proposed (Xiang et al. 2008). This method can be applied conveniently to the cylindrical shells with partial PCLD treatment under various boundary conditions. The state vector is formulated by the displacement and internal force components of the host shell and CL and provides a new means for the analysis of PCLD cylindrical shell.In order to enhance the vibration energy dissipation of the cylindrical shell with conventional PCLD treatment, the influence of multilayer PCLD treatment on vibration and noise suppression of the cylindrical shell loaded with liquid was investigated and analyzed (Saravanan et al. 2000).

This article applied the transfer function method to the dynamic analysis of multi-layer PCLD cylindrical shell. Based on the Donnell assumptions and linear visco-elastic theory, the constitutive relations for the five-layer PCLD cylindrical shell can be described. In terms of energy, the motion equations and boundary conditions are derived for the five-layer PCLD cylindrical shell by Hamilton principle. After trigonometric series expansion and Laplace transform, the state vector is introduced and the motion equation in state space is established. Then the state equation is solved by the transfer function method. Comparing with the traditional three-layer PCLD cylindrical shell, the results reveal that multi-layer PCLD cylindrical shell is more effective than the traditional three-layer PCLD cylindrical shell in suppressing vibration.

2. The constitutive relations

Figure 1 shows the geometry and coordinate system of five-layer PCLD cylindrical shell which was composed of the host shell (layer1), the VEM layers (layer 2 and layer 4) and the constrained layers (layer 3 and layer 5). x coordinate is in the axial direction of the shell, θ and the z coordinates are in the circumferential and radial directions of the shell respectively. R_i (i=1,2,3,4,5) denotes the middle surface radius of each layer; h_i denotes the thickness of each layer; and L denotes the length of the cylindrical shell.

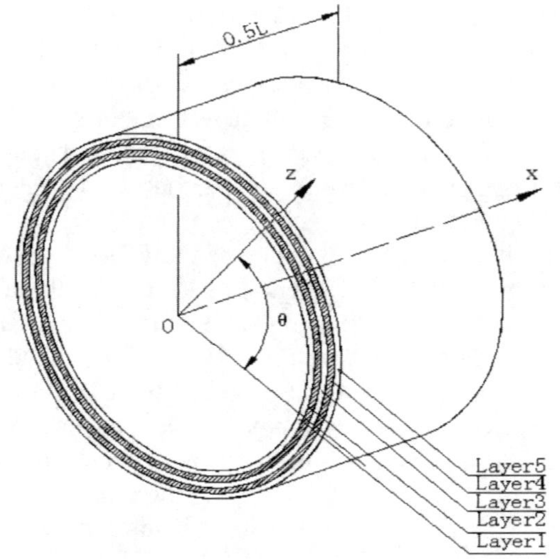

Figure 1. Geometry and coordinate system of five-layer PCLD cylindrical shell

To simplify the computational effort, the following assumptions are proposed (Li et al. 2008): 1) The extrusion deformation in z direction is negligible, that is, all layers have the same displacement in z direction; 2) There is no slip at the interfaces between layers, that is, perfect continuity; 3) The VEM undergoes only shear strains;4) The host shell and constrained layers satisfy Kirchhoff's thin shell hypothesis and their shear strains were ignored; 5) all the cross-section moment of inertias are neglected; 6) all VEM's vibration inertias except the radial were neglected.

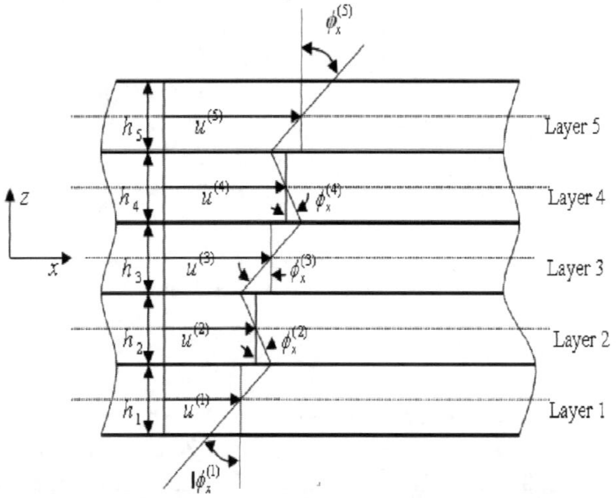

Figure 2. The deformation compatibility between layers

2.1. Displacements

Figure 2 shows the deformation compatibility between layers, the displacements of five-layer PCLD cylindrical shell can be written as

$$\begin{cases} \bar{u}^{(i)} = u^{(i)} + z\phi_x^{(i)} \\ \bar{v}^{(i)} = v^{(i)} + z\phi_\theta^{(i)} \end{cases} \quad i = 1,2,3,4,5 \tag{1}$$

Where the $u^{(i)}$ and $v^{(i)}$ denote the middle surface displacement amplitudes of the ith layer in the x and θ directions respectively; the w is the displacement amplitude in the z direction; the $\phi_x^{(i)}$ and $\phi_\theta^{(i)}$ are rotation angles of the ith layer around the x and θ coordinates respectively.

Based on the thin shell hypothesis, the following relationships are obtained

$$\begin{cases} \phi_x^{(1)} = \phi_x^{(3)} = \phi_x^{(5)} = -\dfrac{\partial w}{\partial x} \\ \phi_\theta^{(1)} = -\dfrac{\partial w}{R_1 \partial\theta} \\ \phi_\theta^{(3)} = -\dfrac{\partial w}{R_3 \partial\theta} \\ \phi_\theta^{(5)} = -\dfrac{\partial w}{R_5 \partial\theta} \end{cases} \tag{2}$$

By using the displacement continuity between layers, the following relationships are yielded

$$\begin{cases} u^{(2)} = \dfrac{1}{2}(u^{(1)} + \dfrac{h_1}{2}\phi_x^{(1)} + u^{(3)} - \dfrac{h_3}{2}\phi_x^{(3)}) \\ \phi_x^{(2)} = \dfrac{1}{h_2}(u^{(3)} - \dfrac{h_3}{2}\phi_x^{(3)} - u^{(1)} - \dfrac{h_1}{2}\phi_x^{(1)}) \\ v^{(2)} = \dfrac{1}{2}(v^{(1)} + \dfrac{h_1}{2}\phi_\theta^{(1)} + v^{(3)} - \dfrac{h_3}{2}\phi_\theta^{(3)}) \\ \phi_\theta^{(2)} = \dfrac{1}{h_2}(v^{(3)} - \dfrac{h_3}{2}\phi_\theta^{(3)} - v^{(1)} - \dfrac{h_1}{2}\phi_\theta^{(1)}) \end{cases} \tag{3}$$

$$\begin{cases} u^{(4)} = \dfrac{1}{2}(u^{(3)} + \dfrac{h_3}{2}\phi_x^{(3)} + u^{(5)} - \dfrac{h_5}{2}\phi_x^{(5)}) \\ \phi_x^{(4)} = \dfrac{1}{h_4}(u^{(5)} - \dfrac{h_5}{2}\phi_x^{(5)} - u^{(3)} - \dfrac{h_3}{2}\phi_x^{(3)} \\ v^{(4)} = \dfrac{1}{2}(v^{(3)} + \dfrac{h_3}{2}\phi_\theta^{(3)} + v^{(5)} - \dfrac{h_5}{2}\phi_\theta^{(5)}) \\ \phi_\theta^{(4)} = \dfrac{1}{h_4}(v^{(4)} - \dfrac{h_5}{2}\phi_\theta^{(5)} - v^{(3)} - \dfrac{h_3}{2}\phi_\theta^{(3)}) \end{cases} \tag{4}$$

2.2. Strains

Based on the Kirchhoff's thin shell hypothesis, the middle surface strains of the host shell and constrained layers can be written as

$$\begin{cases} \varepsilon_{xx}^{o(i)} = \dfrac{\partial u^{(i)}}{\partial x} \\ \varepsilon_{\theta\theta}^{o(i)} = \dfrac{\partial v^{(i)}}{R_i \partial\theta} + \dfrac{w}{R_i} \\ \varepsilon_{x\theta}^{o(i)} = \dfrac{\partial v}{\partial x} + \dfrac{\partial u^{(i)}}{R_i \partial\theta} \end{cases} \quad i = 1,3,5 \tag{5}$$

On the other hand, the middle surface curvatures of the host shell and constrained layers can also be written as

$$\begin{cases} \kappa_{xx}^{(i)} = -\dfrac{\partial^2 w}{\partial x^2} \\ \kappa_{\theta\theta}^{(i)} = -\dfrac{1}{R_i^2}\dfrac{\partial^2 w}{\partial\theta^2} \\ \kappa_{x\theta}^{(i)} = -\dfrac{2}{R_i}\dfrac{\partial^2 w}{\partial x \partial\theta} \end{cases} \quad i = 1,3,5 \tag{6}$$

The strains of the host shell and constrained layers can be further derived

$$\begin{cases} \varepsilon_{xx}^{(i)} = \varepsilon_{xx}^{o(i)} + z\kappa_{xx}^{(i)} \\ \varepsilon_{\theta\theta}^{(i)} = \varepsilon_{\theta\theta}^{o(i)} + z\kappa_{\theta\theta}^{(i)} \\ \varepsilon_{x\theta}^{(i)} = \varepsilon_{x\theta}^{o(i)} + z\kappa_{x\theta}^{(i)} \end{cases} \quad i = 1,3,5 \tag{7}$$

The strains of VEM layers can be obtained

$$\begin{cases} \varepsilon_{xz}^{(2)} = \dfrac{\partial w}{\partial x} + \phi_x^{(2)} \\ \varepsilon_{\theta z}^{(2)} = \dfrac{1}{R_2}\dfrac{\partial w}{\partial \theta} - \dfrac{(v^{(2)} + z\phi_\theta^{(2)})}{R_2} + \phi_\theta^{(2)} \end{cases} \tag{8}$$

$$\begin{cases} \varepsilon_{xz}^{(4)} = \dfrac{\partial w}{\partial x} + \phi_x^{(4)} \\ \varepsilon_{\theta z}^{(4)} = \dfrac{1}{R_4}\dfrac{\partial w}{\partial \theta} - \dfrac{(v^{(4)} + z\phi_\theta^{(4)})}{R_4} + \phi_\theta^{(4)} \end{cases} \tag{9}$$

Substituted Eq.(3) and Eq.(4) into Eq.(8)and Eq.(9), and assuming $\frac{z}{R_2} \approx 0$ and $\frac{z}{R_4} \approx 0$ for the thin shell, the following relationships can be derived

$$\begin{cases} \varepsilon_{xz}^{(2)} = \dfrac{1}{h_2}(u_3 - u_1) + p_{x2}\dfrac{\partial w}{\partial x} \\ \varepsilon_{\theta z}^{(2)} = (\dfrac{1}{h_2} - \dfrac{1}{2R_2})v_3 + (\dfrac{1}{h_2} + \dfrac{1}{2R_2})v_1 \\ \quad + p_{\theta 2}\dfrac{\partial w}{\partial \theta} \end{cases} \tag{10}$$

$$\begin{cases} \varepsilon_{xz}^{(4)} = \dfrac{1}{h_4}(u_5 - u_3) + p_{x4}\dfrac{\partial w}{\partial x} \\ \varepsilon_{\theta z}^{(4)} = (\dfrac{1}{h_4} - \dfrac{1}{2R_4})v_5 + (\dfrac{1}{h_4} + \dfrac{1}{2R_4})v_3 \\ \quad + p_{\theta 4}\dfrac{\partial w}{\partial \theta} \end{cases} \tag{11}$$

Where, some geometric relationships are applied

$$p_{x2} = \frac{1}{2h_2}(h_1 + 2h_2 + h_3)$$

$$p_{\theta 2} = \frac{h_3}{2h_2R_3} + \frac{h_1}{2h_2R_1} + \frac{h_1}{2R_1R_2} - \frac{h_3}{2R_2R_3} + \frac{1}{R_2}$$

$$p_{x4} = \frac{1}{2h_4}(h_3 + 2h_4 + h_5)$$

$$p_{\theta 4} = \frac{h_5}{2h_4R_5} + \frac{h_3}{2h_4R_3} + \frac{h_3}{2R_3R_4} - \frac{h_5}{2R_4R_5} + \frac{1}{R_4}.$$

2.3. Stresses

Using the Hooke's Law, the stresses of the host shell and constrained layers are obtained

$$\begin{bmatrix} \sigma_{xx}^{(i)} \\ \sigma_{\theta\theta}^{(i)} \\ \sigma_{x\theta}^{(i)} \end{bmatrix} = \begin{bmatrix} Q_{11} & Q_{12} & 0 \\ Q_{12} & Q_{22} & 0 \\ 0 & 0 & Q_{66} \end{bmatrix} \begin{bmatrix} \varepsilon_{xx}^{(i)} \\ \varepsilon_{\theta\theta}^{(i)} \\ \varepsilon_{x\theta}^{(i)} \end{bmatrix} \quad i = 1,3,5 \tag{12}$$

Where, $Q_{11} = Q_{22} = \dfrac{E_i}{1-\mu_i^2}$, $Q_{12} = \dfrac{E_i\mu_i}{1-\mu_i^2}$, $Q_{66} = \dfrac{E_i}{2(1+\mu_i)}$, E_i and μ_i denote the elastic modulus and poisson ratio of the ith layer's material respectively. The shear stresses of VEM layers can be written as

$$\sigma_{xz}^{(2)} = G_2\varepsilon_{xz}^{(2)} \quad \sigma_{\theta z}^{(2)} = G_2\varepsilon_{\theta z}^{(2)} \tag{13}$$

$$\sigma_{xz}^{(4)} = G_4\varepsilon_{xz}^{(4)} \quad \sigma_{\theta z}^{(4)} = G_4\varepsilon_{\theta z}^{(4)} \tag{14}$$

Where, G_2 and G_4 are the complex shear modulus of VEM layers in 2nd and 4th layers, respectively.

3. Equations of motion

The strain energy of the host shell can be written as

$$\begin{aligned} U_1 &= \frac{1}{2}\iiint(\sigma_{xx}^{(1)}\varepsilon_{xx}^{(1)} + \sigma_{\theta\theta}^{(1)}\varepsilon_{\theta\theta}^{(1)} + \sigma_{x\theta}^{(1)}\varepsilon_{x\theta}^{(1)})R_1 dz d\theta dx \\ &= \frac{1}{2}\iint K_1[(\varepsilon_{xx}^{o(1)} + \varepsilon_{\theta\theta}^{o(1)})^2 - 2(1-\mu_1)(\varepsilon_{xx}^{o(1)}\varepsilon_{\theta\theta}^{o(1)} - \\ &\quad \frac{\varepsilon_{x\theta}^{o(1)2}}{4})]R_1 d\theta dx + \frac{1}{2}\iint D_1[(\kappa_{xx}^{(1)} + \kappa_{\theta\theta}^{(1)})^2 \\ &\quad - 2(1-\mu_1)(\kappa_{xx}^{(1)}\kappa_{\theta\theta}^{(1)} - \frac{\kappa_{x\theta}^{(1)2}}{4})]R_1 d\theta dx \end{aligned} \tag{15}$$

where, $K_1 = \dfrac{E_1 h_1}{1-\mu_1^2}$, $D_1 = \dfrac{E_1 h_1^3}{12(1-\mu_1^2)}$. Replacing the subscript "1" with "3" and "5", the strain energies of two constrained layers U_3 and U_5 can be written in the same formula as Eq.(15) respectively.

The strain energies of the VEM layers can be written as

$$\begin{aligned} U_2 &= \frac{1}{2}\iiint(\sigma_{xz}^{(2)}\varepsilon_{xz}^{(2)} + \sigma_{\theta z}^{(2)}\varepsilon_{\theta z}^{(2)})R_2 dz d\theta dx \\ &= \frac{R_2 G_2 h_2}{2}\iint(\varepsilon_{xz}^{(2)2} + \varepsilon_{\theta z}^{(2)2})d\theta dx \end{aligned} \tag{16}$$

$$\begin{aligned} U_4 &= \frac{1}{2}\iiint(\sigma_{xz}^{(4)}\varepsilon_{xz}^{(4)} + \sigma_{\theta z}^{(4)}\varepsilon_{\theta z}^{(4)})R_4 dz d\theta dx \\ &= \frac{R_4 G_4 h_4}{2}\iint(\varepsilon_{xz}^{(4)2} + \varepsilon_{\theta z}^{(4)2})d\theta dx \end{aligned} \tag{17}$$

Therefore, the total strain energy of five-layer PCLD cylindrical shell is expressed by

$$U = U_1 + U_2 + U_3 + U_4 + U_5 \tag{18}$$

Taking the assumption (6) into consideration, the total kinetic energy is generated

$$\begin{aligned} T = \frac{1}{2}\iint[& m_1 R_1(\dot{u}_1^2 + \dot{v}_1^2 + \dot{w}^2) + m_2 R_2 \dot{w}^2 + \\ & m_3 R_3(\dot{u}_3^2 + \dot{v}_3^2 + \dot{w}^2) + m_4 R_4 \dot{w}^2 + \\ & m_5 R_5(\dot{u}_5^2 + \dot{v}_5^2 + \dot{w}^2)]d\theta dx \end{aligned} \tag{19}$$

If the distributed load $q(x,\theta,t)$ is applied on the host shell, the virtual work from the distributed load is yield

$$\delta W = R_1 \iint q(x,\theta,t)\delta w d\theta dx \tag{20}$$

According to the Hamilton's principle: $\int_{t_0}^{t_1}(\delta U - \delta W - \delta T)dt = 0$, simplifying the motion equations by $\dfrac{R_j}{R_i} \approx 1$, $1 \pm \dfrac{h_j}{2R_i} \approx 1$, then the motion equations of five-layer PCLD cylindrical shell can be derived

$$
\begin{aligned}
& m_1\ddot{u}_1 - K_1\left(\frac{\partial^2 u_1}{\partial x^2} + \frac{1+\mu_1}{2R_1}\frac{\partial^2 v_1}{\partial\theta\partial x} + \frac{\mu_1}{R_1}\frac{\partial w}{\partial x} + \frac{1-\mu_1}{2R_1^2}\frac{\partial^2 u_1}{\partial\theta^2}\right) - G_2\varepsilon_{xz}^{(2)} = 0 \\
& m_1\ddot{v}_1 - K_1\left(\frac{1}{R_1^2}\frac{\partial^2 v_1}{\partial\theta^2} + \frac{1+\mu_1}{2R_1}\frac{\partial^2 u_1}{\partial\theta\partial x} + \frac{1}{R_1^2}\frac{\partial w}{\partial\theta} + \frac{1-\mu_1}{2}\frac{\partial^2 v_1}{\partial x^2}\right) - G_2\varepsilon_{\theta z}^{(2)} = 0 \\
& m_3\ddot{u}_3 - K_3\left(\frac{\partial^2 u_3}{\partial x^2} + \frac{1+\mu_3}{2R_3}\frac{\partial^2 v_3}{\partial\theta\partial x} + \frac{\mu_3}{R_3}\frac{\partial w}{\partial x} + \frac{1-\mu_3}{2R_3^2}\frac{\partial^2 u_3}{\partial\theta^2}\right) + G_2\varepsilon_{xz}^{(2)} - G_4\varepsilon_{xz}^{(4)} = 0 \\
& m_3\ddot{v}_3 - K_3\left(\frac{1}{R_3^2}\frac{\partial^2 v_3}{\partial\theta^2} + \frac{1+\mu_3}{2R_3}\frac{\partial^2 u_3}{\partial\theta\partial x} + \frac{1}{R_3^2}\frac{\partial w}{\partial\theta} + \frac{1-\mu_3}{2}\frac{\partial^2 v_3}{\partial x^2}\right) + G_2\varepsilon_{\theta z}^{(2)} - G_4\varepsilon_{\theta z}^{(4)} = 0 \\
& m_5\ddot{u}_5 - K_5\left(\frac{\partial^2 u_5}{\partial x^2} + \frac{1+\mu_5}{2R_5}\frac{\partial^2 v_5}{\partial\theta\partial x} + \frac{\mu_5}{R_5}\frac{\partial w}{\partial x} + \frac{1-\mu_5}{2R_5^2}\frac{\partial^2 u_5}{\partial\theta^2}\right) + G_4\varepsilon_{xz}^{(4)} = 0 \\
& m_5\ddot{v}_5 - K_5\left(\frac{1}{R_5^2}\frac{\partial^2 v_5}{\partial\theta^2} + \frac{1+\mu_5}{2R_5}\frac{\partial^2 u_5}{\partial\theta\partial x} + \frac{1}{R_5^2}\frac{\partial w}{\partial\theta} + \frac{1-\mu_5}{2}\frac{\partial^2 v_5}{\partial x^2}\right) + G_4\varepsilon_{\theta z}^{(4)} = 0 \\
& m\ddot{w} + K_1\left(\frac{w}{R_1^2} + \frac{1}{R_1^2}\frac{\partial v_1}{\partial\theta} + \frac{\mu_1}{R_1}\frac{\partial u_1}{\partial x}\right) + K_3\left(\frac{w}{R_3^2} + \frac{1}{R_3^2}\frac{\partial v_3}{\partial\theta} + \frac{\mu_3}{R_3}\frac{\partial u_3}{\partial x}\right) + \\
& K_5\left(\frac{w}{R_5^2} + \frac{1}{R_5^2}\frac{\partial v_5}{\partial\theta} + \frac{\mu_5}{R_5}\frac{\partial u_5}{\partial x}\right) + D_1\left(\frac{\partial^4 w}{\partial x^4} + \frac{1}{R_1^2}\frac{\partial^4 w}{\partial\theta^2\partial x^2} + \frac{1}{R_1^4}\frac{\partial^4 w}{\partial\theta^4}\right) + \\
& D_3\left(\frac{\partial^4 w}{\partial x^4} + \frac{1}{R_3^2}\frac{\partial^4 w}{\partial\theta^2\partial x^2} + \frac{1}{R_3^4}\frac{\partial^4 w}{\partial\theta^4}\right) + D_5\left(\frac{\partial^4 w}{\partial x^4} + \frac{1}{R_5^2}\frac{\partial^4 w}{\partial\theta^2\partial x^2} + \frac{1}{R_5^4}\frac{\partial^4 w}{\partial\theta^4}\right) - \\
& G_2h_2\left(p_{x2}\frac{\partial\varepsilon_{xz}^{(2)}}{\partial x} + p_{\theta 2}\frac{\partial\varepsilon_{\theta z}^{(2)}}{\partial\theta}\right) - G_4h_4\left(p_{x4}\frac{\partial\varepsilon_{xz}^{(4)}}{\partial x} + p_{\theta 4}\frac{\partial\varepsilon_{\theta z}^{(4)}}{\partial\theta}\right) - q(x,\theta,t) = 0
\end{aligned}
\tag{21}
$$

Here, $m=m_1+m_2+m_3+m_4+m_5$. Boundary conditions of clamped-clamped cylindrical shell at $x=-0.5L$ and $x=0.5L$ are further given by

$$
\begin{aligned}
& u_1 = 0 \quad or \quad N_{x1} = K_1\left(\frac{\partial u_1}{\partial x} + \frac{\mu_1}{R_1}\frac{\partial v_1}{\partial\theta} + \frac{\mu_1}{R_1}w\right) = 0 \\
& v_1 = 0 \quad or \quad N_{x\theta 1} = K_1\frac{1-\mu_1}{2}\left(\frac{\partial v_1}{\partial x} + \frac{1}{R_1}\frac{\partial u_1}{\partial\theta}\right) = 0 \\
& u_3 = 0 \quad or \quad N_{x3} = K_3\left(\frac{\partial u_3}{\partial x} + \frac{\mu_3}{R_3}\frac{\partial v_3}{\partial\theta} + \frac{\mu_3}{R_3}w\right) = 0 \\
& v_3 = 0 \quad or \quad N_{x\theta 3} = K_3\frac{1-\mu_3}{2}\left(\frac{\partial v_3}{\partial x} + \frac{1}{R_3}\frac{\partial u_3}{\partial\theta}\right) = 0 \\
& u_5 = 0 \quad or \quad N_{x5} = K_5\left(\frac{\partial u_5}{\partial x} + \frac{\mu_5}{R_5}\frac{\partial v_5}{\partial\theta} + \frac{\mu_5}{R_5}w\right) = 0 \\
& v_5 = 0 \quad or \quad N_{x\theta 5} = K_5\frac{1-\mu_5}{2}\left(\frac{\partial v_5}{\partial x} + \frac{1}{R_5}\frac{\partial u_5}{\partial\theta}\right) = 0 \\
& \frac{\partial w}{\partial x} = 0 \quad or \quad M_x = D_1\left(\frac{\partial^2 w}{\partial x^2} + \frac{\mu_1}{R_1^2}\frac{\partial^2 w}{\partial\theta^2}\right) + D_3\left(\frac{\partial^2 w}{\partial x^2}\right. \\
& \qquad \left. + \frac{\mu_3}{R_3^2}\frac{\partial^2 w}{\partial\theta^2}\right) + D_5\left(\frac{\partial^2 w}{\partial x^2} + \frac{\mu_5}{R_5^2}\frac{\partial^2 w}{\partial\theta^2}\right) = 0 \\
& w = 0 \quad or \quad Q_x = D_1\left(\frac{\partial^3 w}{\partial x^3} + \frac{2-\mu_1}{R_1^2}\frac{\partial^3 w}{\partial x\partial\theta^2}\right) + D_3\left(\frac{\partial^3 w}{\partial x^3}\right. \\
& \qquad \left. + \frac{2-\mu_3}{R_3^2}\frac{\partial^3 w}{\partial x\partial\theta^2}\right) + D_5\left(\frac{\partial^3 w}{\partial x^3} + \frac{2-\mu_5}{R_5^2}\frac{\partial^3 w}{\partial x\partial\theta^2}\right) \\
& \qquad - G_2h_2p_{x2}\varepsilon_{xz}^{(2)} - G_4h_4p_{x4}\varepsilon_{xz}^{(4)} = 0
\end{aligned}
\tag{22}
$$

For the isotropic PCLD cylindrical shells, the following formulas based on Fourier series are introduced to simplify the computed process

$$
\begin{aligned}
& u_1 = \sum_{n=0}^{\infty} U_{1n}(x,t)\cos n\theta \quad v_1 = \sum_{n=0}^{\infty} V_{1n}(x,t)\sin n\theta \\
& u_3 = \sum_{n=0}^{\infty} U_{3n}(x,t)\cos n\theta \quad v_3 = \sum_{n=0}^{\infty} V_{3n}(x,t)\sin n\theta \\
& u_5 = \sum_{n=0}^{\infty} U_{5n}(x,t)\cos n\theta \quad v_5 = \sum_{n=0}^{\infty} V_{5n}(x,t)\sin n\theta \\
& w = \sum_{n=0}^{\infty} W_n(x,t)\cos n\theta \quad q = \sum_{n=0}^{\infty} q_n(x,t)\cos n\theta
\end{aligned}
\tag{23}
$$

where n denotes the circumferential wave number.

Substituting Eq.(23) into Eq.(21), and then Laplace transformation is made for Eq.(21). The following state equations in Laplace domain are yielded

$$
\begin{aligned}
\frac{\partial^2\tilde{U}_{1n}}{\partial x^2} &= A^{u_1}_{u_1^0}\tilde{U}_{1n} + A^{u_1}_{v_1^1}\frac{\partial\tilde{V}_{1n}}{\partial x} + A^{u_1}_{u_3^0}\tilde{U}_{3n} + A^{u_1}_{w^1}\frac{\partial\tilde{W}_n}{\partial x}\\
\frac{\partial^2\tilde{V}_{1n}}{\partial x^2} &= A^{v_1}_{u_1^1}\frac{\partial\tilde{U}_{1n}}{\partial x} + A^{v_1}_{v_1^0}\tilde{V}_{1n} + A^{v_1}_{v_3^0}\tilde{V}_{3n} + A^{v_1}_{w^0}\tilde{W}_n\\
\frac{\partial^2\tilde{U}_{3n}}{\partial x^2} &= A^{u_3}_{u_1^0}\tilde{U}_{1n} + A^{u_3}_{u_3^0}\tilde{U}_{3n} + A^{u_3}_{v_3^1}\frac{\partial\tilde{V}_{3n}}{\partial x} + A^{u_3}_{u_5^0}\tilde{U}_{5n} + A^{u_3}_{w^1}\frac{\partial\tilde{W}_n}{\partial x}\\
\frac{\partial^2\tilde{V}_{3n}}{\partial x^2} &= A^{v_3}_{v_1^0}\tilde{V}_{1n} + A^{v_3}_{u_3^1}\frac{\partial\tilde{U}_{3n}}{\partial x} + A^{v_3}_{v_3^0}\tilde{V}_{3n} + A^{V_3}_{v_5^0}V_{5n} + A^{v_3}_{w^0}\tilde{W}_n\\
\frac{\partial^2\tilde{U}_{5n}}{\partial x^2} &= A^{u_5}_{u_3^0}\tilde{U}_{3n} + A^{u_5}_{u_5^0}\tilde{U}_{5n} + A^{u_5}_{v_5^1}\frac{\partial\tilde{V}_{5n}}{\partial x} + A^{u_5}_{w^1}\frac{\partial\tilde{W}_n}{\partial x}\\
\frac{\partial^2\tilde{V}_{5n}}{\partial x^2} &= A^{v_5}_{v_3^0}\tilde{V}_{3n} + A^{v_5}_{u_5^1}\frac{\partial\tilde{U}_{5n}}{\partial x} + A^{v_5}_{v_5^0}\tilde{V}_{5n} + A^{v_5}_{w^0}\tilde{W}_n\\
\frac{\partial^4\tilde{W}_n}{\partial x^4} &= A^{w}_{u_1^1}\frac{\partial\tilde{U}_{1n}}{\partial x} + A^{w}_{v_1^0}\tilde{V}_{1n} + A^{w}_{u_3^1}\frac{\partial\tilde{U}_{3n}}{\partial x} + A^{w}_{v_3^0}\tilde{V}_{3n} + A^{w}_{u_5^1}\frac{\partial\tilde{U}_{5n}}{\partial x}\\
&\quad + A^{w}_{v_5^0}\tilde{V}_{5n} + A^{w}_{w^0}\tilde{W}_n + A^{w}_{w^2}\frac{\partial^2\tilde{W}_n}{\partial x^2} + \tilde{f}^{w}
\end{aligned}
\tag{24}
$$

All relevant coefficients in Eq.(24) are presented in appendix A.

Similarly, the boundary condition in Laplace domain can be derived

$$
\begin{aligned}
&\tilde{U}_{1n}=0 \quad or \quad \tilde{N}_{x1}=K_1\left(\frac{\partial\tilde{U}_{1n}}{\partial x}+\frac{\mu_1}{R_1}n\tilde{V}_{1n}+\frac{\mu_1}{R_1}\tilde{W}_n\right)=0\\
&\tilde{V}_{1n}=0 \quad or \quad \tilde{N}_{x\theta1}=K_1\frac{1-\mu_1}{2}\left(\frac{\partial\tilde{V}_{1n}}{\partial x}-\frac{n}{R_1}\tilde{U}_{1n}\right)=0\\
&\tilde{U}_{3n}=0 \quad or \quad \tilde{N}_{x3}=K_3\left(\frac{\partial\tilde{U}_{3n}}{\partial x}+\frac{\mu_3}{R_3}n\tilde{V}_{3n}+\frac{\mu_3}{R_3}\tilde{W}_n\right)=0\\
&\tilde{V}_{3n}=0 \quad or \quad \tilde{N}_{x\theta3}=K_3\frac{1-\mu_3}{2}\left(\frac{\partial\tilde{V}_{3n}}{\partial x}-\frac{n}{R_3}\tilde{U}_{3n}\right)=0\\
&\tilde{U}_{5n}=0 \quad or \quad \tilde{N}_{x5}=K_5\left(\frac{\partial\tilde{U}_{5n}}{\partial x}+\frac{\mu_5}{R_5}n\tilde{V}_{5n}+\frac{\mu_5}{R_5}\tilde{W}_n\right)=0\\
&\tilde{V}_{5n}=0 \quad or \quad \tilde{N}_{x\theta5}=K_5\frac{1-\mu_5}{2}\left(\frac{\partial\tilde{V}_{5n}}{\partial x}-\frac{n}{R_5}\tilde{U}_{5n}\right)=0\\
&\frac{\partial\tilde{W}_n}{\partial x}=0 \quad or \quad \tilde{M}_x=(D_1+D_3+D_5)\frac{\partial^2\tilde{W}_n}{\partial x^2}-\left(D_1\frac{\mu_1}{R_1^2}\right.\\
&\left.+D_3\frac{\mu_3}{R_3^2}+D_5\frac{\mu_5}{R_5^2}\right)n^2\tilde{W}_n=0\\
&\tilde{W}_n=0 \quad or \quad \tilde{Q}_x=(D_1+D_3+D_5)\frac{\partial^3\tilde{W}_n}{\partial x^3}-\left[\frac{n^2D_1(2-\mu_1)}{R_1^2}\right.\\
&+\frac{n^2D_3(2-\mu_3)}{R_3^2}+\frac{n^2D_5(2-\mu_5)}{R_5^2}+G_2h_2p_{x2}^2+G_4h_4p
\end{aligned}
\tag{25}
$$

4. Transfer function method

In order to solve Eq.(24), the state vector is defined as follows

$$
\begin{aligned}
\tilde{\beta}_n=\{&\tilde{U}_{1n} \quad \frac{\partial\tilde{U}_{1n}}{\partial x} \quad \tilde{V}_{1n} \quad \frac{\partial\tilde{V}_{1n}}{\partial x} \quad \tilde{U}_{3n} \quad \frac{\partial\tilde{U}_{3n}}{\partial x} \quad \tilde{V}_{3n} \quad \frac{\partial\tilde{V}_{3n}}{\partial x} \quad \tilde{U}_{5n}\\
&\frac{\partial\tilde{U}_{5n}}{\partial x} \quad \tilde{V}_{5n} \quad \frac{\partial\tilde{V}_{5n}}{\partial x} \quad \tilde{W}_n \quad \frac{\partial\tilde{W}}{\partial x} \quad \frac{\partial^2\tilde{W}}{\partial x^2} \quad \frac{\partial^3\tilde{W}}{\partial x^3}\}^T
\end{aligned}
$$

In this way, Eq.(24) can be written as

$$
\frac{\partial}{\partial x}\tilde{\beta}_n(x,s)=B_n(s)\tilde{\beta}_n(x,s)+\tilde{f}_n(x,s) \tag{26}
$$

Where, $B_n(s)$ and $\tilde{f}_n(x,s)$ are presented in appendix B.

For clamped-clamped five-layer PCLD cylindrical shell, the boundary conditions can be expressed by

$$P_n\tilde{\beta}_n(-0.5L,s)+Q_n\tilde{\beta}_n(0.5L,s)=0 \tag{27}$$

Where, $P_n=\begin{bmatrix}\Lambda\\ \hline 0_{8\times16}\end{bmatrix}$, $Q_n=\begin{bmatrix}0_{8\times16}\\ \hline \Lambda\end{bmatrix}$,

$$\Lambda=\begin{bmatrix}1&0&0&0&0&0&0&0&0&0&0&0&0&0&0&0\\0&0&1&0&0&0&0&0&0&0&0&0&0&0&0&0\\0&0&0&0&1&0&0&0&0&0&0&0&0&0&0&0\\0&0&0&0&0&0&1&0&0&0&0&0&0&0&0&0\\0&0&0&0&0&0&0&0&1&0&0&0&0&0&0&0\\0&0&0&0&0&0&0&0&0&0&1&0&0&0&0&0\\0&0&0&0&0&0&0&0&0&0&0&0&1&0&0&0\\0&0&0&0&0&0&0&0&0&0&0&0&0&1&0&0\end{bmatrix}.$$

The solution of Eq.(26) can be written as

$$\tilde{\beta}_n=\int_{-0.5L}^{0.5L}D_n(x,\xi,s)\tilde{f}_n(\xi,s)d\xi \tag{28}$$

Where, $$D_n(x,\xi,s)=\begin{cases}e^{B_n(s)}[P_ne^{-0.5LB_n(s)}+Q_ne^{0.5LB_n(s)}]^{-1}P_ne^{B_n(s)(-0.5L-\xi)},\xi\le x\\-e^{B_n(s)}[P_ne^{-0.5LB_n(s)}+Q_ne^{0.5LB_n(s)}]^{-1}Q_ne^{B_n(s)(0.5L-\xi)},\xi>x\end{cases}.$$

By solving the characteristic equation

$$\det[P_ne^{-0.5LB_n(s)}+Q_ne^{0.5LB_n(s)}]=0 \tag{29}$$

natural frequency and loss factor of five-layer PCLD cylindrical shell can be obtained. All the derived processes and solving approach for the five-layer PCLD cylindrical shell can be extended to seven-layer and even more layer PCLD cylindrical shells.

5. Numerical results and analysis

5.1. The five-layer PCLD cylindrical shell

The geometric and physical properties of five-layer PCLD cylindrical shell are presented as following: R_1=0.3m, L=0.1m, h_1=0.003m, h_2=0.0005m, h_3=0.001m, h_4=0.0005m, h_5=0.001m, G_2=G_4=0.896(1+0.9683i)Mpa, $\mu_1=\mu_3=\mu_5=0.3$, $\mu_2=\mu_4=0.498$, $\rho_1=\rho_3=\rho_5=2700kg/m^3$ $\rho_2=\rho_4=999kg/m^3$.

The boundary condition is the clamped-clamped.

Table 1. Comparison of TFM with ANSYS for five-layer PCLD cylindrical shell

	Natural frequency (Hz)			Loss factor (%)		
N	TFM	ANSYS	error (%)	TFM	ANSYS	error (%)
0	2975	2976	0.034	0.357	0.359	0.560
1	2942	2943	0.034	0.367	0.369	0.545
2	2850	2851	0.035	0.398	0.398	0.000
3	2723	2725	0.073	0.450	0.448	0.444
4	2588	2589	0.039	0.523	0.520	0.574
5	2458	2461	0.122	0.617	0.612	0.810
6	2350	2353	0.128	0.728	0.721	0.962
7	2265	2270	0.221	0.851	0.844	0.822
8	2208	2214	0.272	0.980	0.972	0.816
9	2180	2186	0 275	1.104	1.099	0.453
10	2180	2186	0.275	1.216	1.214	0.164

Table 1 shows the natural frequency and loss factor calculated by the proposed transfer function method and commercial finite element analysis software ANSYS. The maximum error of natural frequency and loss factor are 0.275% and 0.962% respectively. Compared to ANSYS approach, the transfer function method needs less process variables and computational cost.

5.2. Multilayer PCLD cylindrical shell

The above five-layer PCLD cylindrical shell can be extended to multilayer PCLD cylindrical shell, while keeping the thickness of the host shell and the total thickness of VEM layers and the total thickness of constrained layers unchanged such as seven-layer, nine-layer and even more layers PCLD cylindrical shell.

Table 2. Natural frequency and loss factor for multilayer PCLD cylindrical shell

	Natural frequency (Hz)					Loss factor (%)				
n	3	5	7	9	11	3	5	7	9	11
0	3029	2975	2966	2964	2963	0.249	0.357	0.422	0.475	0.520
1	2996	2941	2930	2930	2929	0.256	0.367	0.435	0.488	0.535
2	2907	2849	2840	2837	2836	0.276	0.398	0.470	0.529	0.579
3	2784	2723	2712	2710	2709	0.311	0.450	0.532	0.597	0.654
4	2658	2586	2575	2572	2570	0.360	0.523	0.618	0.694	0.759
5	2532	2458	2445	2441	2440	0.421	0.617	0.730	0.819	0.896
6	2431	2348	2334	2330	2329	0.492	0.728	0.862	0.967	1.056
7	2357	2264	2249	2444	2242	0.569	0.851	1.010	1.134	1.250
8	2311	2208	2190	2185	2183	0.648	0.980	1.165	1.309	1.430
9	2295	2180	2160	2154	2150	0.722	1.104	1.317	1.481	1.618
10	2307	2179	2157	2150	2148	0.785	1.216	1.454	1.637	1.790
11	2347	2206	2181	2173	2171	0.835	1.307	1.568	1.767	1.935
12	2415	2258	1230	2222	2219	0.869	1.373	1.651	1.864	2.043
13	2508	2335	2305	2295	2292	0.886	1.412	1.702	1.924	2.110
14	2625	2436	2402	2392	2388	0.887	1.425	1.722	1.948	2.140
15	2766	2559	2522	2510	2506	0.876	1.416	1.713	1.941	2.135
16	2928	2703	2662	2649	2645	0.854	1.388	1.683	1.909	2.101
17	3111	2867	2882	2820	2803	0.826	1.347	1.635	1.850	2.046
18	3315	3049	3000	2996	2979	0.792	1.296	1.575	1.766	1.974
19	3537	3250	3197	3197	3173	0.755	1.240	1.508	1.698	1.894
20	3777	3467	3410	3390	3384	0.718	1.180	1.437	1.630	1.807

Table 2 shows the natural frequency and loss factor for multilayer PCLD cylindrical shell by using transfer function method. It is noted that the natural frequency decreases slightly while the loss factor increases obviously when the layer number is increased, it is special obvious when the circumferential wave number is around 13. Furthermore, when more PCLD layers are bonded to the surface of the host shell while keeping the total thickness of VEM and the total thickness of CL unchanged, the natural frequency decreases or the loss factor increases in a smaller amount with the total layer number increasing.

In order to check the vibration response of multilayer PCLD cylindrical shell, a distributed force $q(x,\theta,t)=\delta(t)\cos(n\theta)Pa$ is applied on the inner side of host shell. Based on the orthogonality of trigonometric function, the steady-state response is sinusoidal with n waves. In this article, we only compute the radial vibration at point (0, 0, 0.3). Figure 3-Figure 5 show the displacement amplitude-frequency response curves for multilayer PCLD cylindrical shell when wave number n is 6, 14 and 20. It is obvious that when the total layer number increases, the peak of curve decreases significantly. The natural frequency around the peak is corresponding with the natural frequency shown in table 2 when the circumferential wave number n is 6, 14, 20 respectively. On the other hand, when the total layer number increases, the natural frequency around the peak decreases rapidly at the beginning and slowly at the end. That means more layers of PCLD will result in much less decrease in the natural frequency of multilayer PCLD cylindrical shell. Meanwhile, it is further proved theoretically that multilayer

PCLD treatment will increase the loss factor significantly, thus enhance the vibration energy dissipation and suppress the vibration and noise of structures. It supplies an effective means to suppress the vibration and noise of structures without extra VEM and CL material.

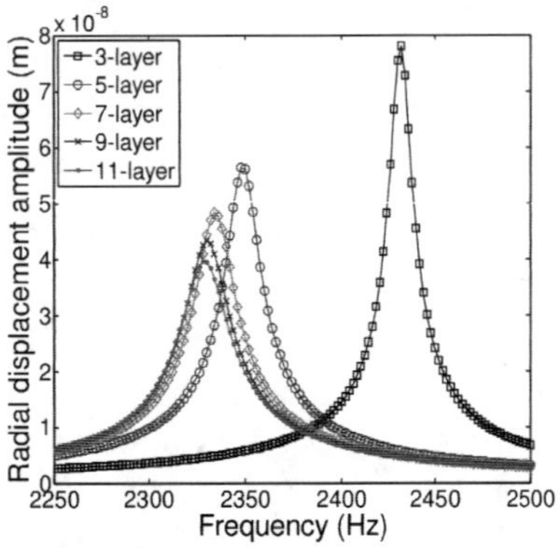

Figure 3. Displacement frequency response for n=6

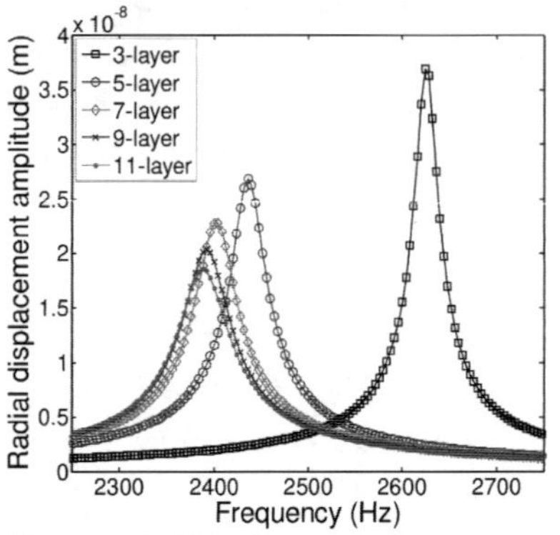

Figure 4. Displacement frequency response for n=14

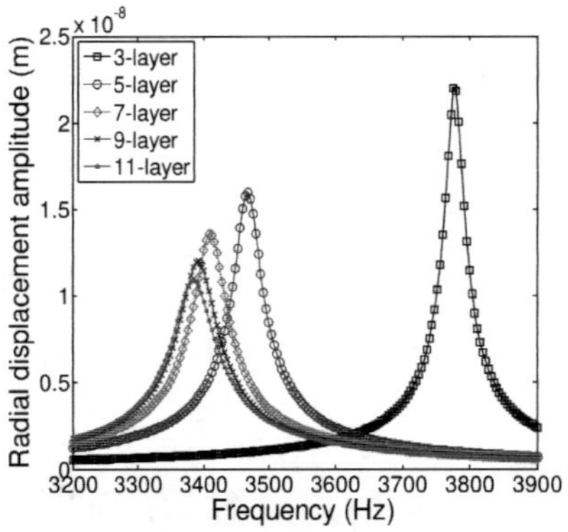

Figure 5. Displacement frequency response for n=20

6. Conclusions

The constitutive relations of the cylindrical shell with five-layer PCLD treatment are described based on the Donnell theory of thin shell. The Hamilton principle is used to derive the motion equation of the cylindrical shell with five-layer PCLD treatment. The state vectors are defined and the state space form of the motion equation is further constructed to solve the motion equation conveniently. The transfer matrix method is applied to solve the state space equation. The natural frequency and the loss factor of the cylindrical shell with conventional three-layer and multilayer PCLD treatments are calculated. The results are concluded as follows:

(1) The application of multilayer PCLD treatment results in a slight drop of natural frequency and a rapid increase in loss factor. Therefore, it supplies an effective means to enhance vibration energy dissipation of the cylindrical shell without extra VEM and CL material applied.

(2) The theoretical analysis in this paper is an important fundamental to push forward the application of multilayer PCLD treatments as a practical, economical and effective means in structural vibration suppression. However, there are still more works to do in this topic, such as the optimization of the layer number, the selection of parameters as well as the layout of PCLD patches. Therefore, the multi-objective optimization will be further considered as an effective design tool to maximize vibration energy dissipation of the cylindrical shell with multilayer PCLD treatment.

Appendix A

$$A^{u_1}_{u_1^0} = [\frac{K_1 n^2(1-\mu_1)}{2R_1^2} + \frac{G_2}{h_2} + m_1 s^2]/K_1, \quad A^{u_1}_{v_1^1} = [-\frac{K_1 n(1+\mu_1)}{2R_1}]/K_1, \quad A^{u_1}_{u_3^0} = (-\frac{G_2}{h_2})/K_1,$$

$$A^{u_1}_{w^1} = -(\frac{K_1\mu_1}{R_1} + G_2 p_x)/K_1, \quad A^{v_1}_{u_1^1} = \frac{K_1 n(1+\mu_1)}{2R_1}/[\frac{K_1(1-\mu_1)}{2}],$$

$$A^{v_1}_{v_1^0} = [\frac{K_1 n^2}{R_1^2} + G_2(\frac{1}{h_2} + \frac{1}{2R_2}) + m_1 s^2]/[\frac{K_1(1-\mu_1)}{2}], \quad A^{v_1}_{v_3^0} = -G_2(\frac{1}{h_2} - \frac{1}{2R_2})/[\frac{K_1(1-\mu_1)}{2}],$$

$$A^{v_1}_{w^0} = (\frac{nK_1}{R_1^2} + G_2 p_\theta n)/[\frac{K_1(1-\mu_1)}{2}], \quad A^{u_3}_{u_1^0} = -(\frac{G_2}{h_2})/K_3, \quad A^{u_3}_{v_3^1} = [-\frac{K_3 n(1+\mu_3)}{2R_3}]/K_3,$$

$$A^{u_3}_{u_3^0} = [\frac{K_3 n^2(1-\mu_3)}{2R_3^2} + \frac{G_2}{h_2} + m_3 s^2]/K_3, \quad A^{u_3}_{w^1} = (G_2 p_x - \frac{K_3\mu_3}{R_3})/K_3, \quad A^{v_3}_{v_1^0} = -G_2(\frac{1}{h_2} + \frac{1}{2R_2})/[\frac{K_3(1-\mu_3)}{2}],$$

$$A^{v_3}_{u_3^1} = \frac{K_3 n(1+\mu_3)}{2R_3}/[\frac{K_3(1-\mu_3)}{2}], \quad A^{v_3}_{v_3^0} = [\frac{K_3 n^2}{R_3^2} + G_2(\frac{1}{h_2} - \frac{1}{2R_2}) + m_3 s^2]/[\frac{K_3(1-\mu_3)}{2}],$$

$$A^{v_3}_{w^0} = (\frac{nK_3}{R_3^2} - G_2 p_\theta n)/[\frac{K_3(1-\mu_3)}{2}], \quad A^{w}_{u_1^1} = -(\frac{K_1\mu_1}{R_1} + G_2 p_x)/(D_1 + D_3),$$

$$A^{w}_{v_1^0} = -(\frac{K_1 n}{R_1^2} + G_2 p_\theta n)/(D_1 + D_3), \quad A^{w}_{u_3^1} = -(\frac{K_3\mu_3}{R_3} - G_2 p_x)/(D_1 + D_3), \quad A^{w}_{v_3^0} = -(\frac{K_3 n}{R_3^2} - G_2 p_\theta n)/(D_1 + D_3),$$

$$A^{w}_{w^0} = -[(\frac{D_1}{R_1^4} + \frac{D_3}{R_3^4})n^4 + \frac{K_1}{R_1^2} + \frac{K_3}{R_3^2} + G_2 h_2 p_\theta^2 n^2 + ms^2]/(D_1 + D_3),$$

$$A^{w}_{w^2} = [2(\frac{D_1}{R_1^4} + \frac{D_3}{R_3^4})n^2 + G_2 h_2 p_x^2]/(D_1 + D_3), \quad \tilde{f}^{w} = \tilde{q}_n/(D_1 + D_3 + D_5)$$

Appendix B

$$B_n(s) = \begin{bmatrix}
0 & 1 & 0 & 0 & 0 & 0 & 0 & 0 & 0 & 0 & 0 & 0 & 0 & 0 & 0 & 0 \\
A^{u_1}_{u_1^0} & 0 & 0 & A^{u_1}_{v_1^1} & A^{u_1}_{u_3^0} & 0 & 0 & 0 & 0 & 0 & 0 & 0 & 0 & A^{u_1}_{w^1} & 0 & 0 \\
0 & 0 & 0 & 1 & 0 & 0 & 0 & 0 & 0 & 0 & 0 & 0 & 0 & 0 & 0 & 0 \\
0 & A^{v_1}_{u_1^1} & A^{v_1}_{v_1^0} & 0 & 0 & 0 & A^{v_1}_{v_3^0} & 0 & 0 & 0 & 0 & 0 & A^{v_1}_{w^0} & 0 & 0 & 0 \\
0 & 0 & 0 & 0 & 0 & 1 & 0 & 0 & 0 & 0 & 0 & 0 & 0 & 0 & 0 & 0 \\
A^{u_3}_{u_1^0} & 0 & 0 & 0 & A^{u_3}_{u_3^0} & 0 & 0 & A^{u_3}_{v_3^1} & A^{u_3}_{u_5^0} & 0 & 0 & 0 & 0 & A^{u_3}_{w^1} & 0 & 0 \\
0 & 0 & 0 & 0 & 0 & 0 & 0 & 1 & 0 & 0 & 0 & 0 & 0 & 0 & 0 & 0 \\
0 & 0 & A^{v_3}_{v_1^0} & 0 & 0 & A^{v_3}_{u_3^1} & A^{v_3}_{v_3^0} & 0 & 0 & 0 & A^{v_3}_{v_5^0} & 0 & A^{v_3}_{w^0} & 0 & 0 & 0 \\
0 & 0 & 0 & 0 & 0 & 0 & 0 & 0 & 0 & 1 & 0 & 0 & 0 & 0 & 0 & 0 \\
0 & 0 & 0 & 0 & A^{U_3}_{U_3^0} & 0 & 0 & 0 & A^{U_3}_{U_5^0} & 0 & 0 & A^{U_3}_{v_5^1} & 0 & A^{U_3}_{w^1} & 0 & 0 \\
0 & 0 & 0 & 0 & 0 & 0 & 0 & 0 & 0 & 0 & 0 & 1 & 0 & 0 & 0 & 0 \\
0 & 0 & 0 & 0 & 0 & 0 & A^{v_5}_{v_3^0} & 0 & 0 & A^{v_5}_{u_5^1} & A^{v_5}_{v_5^0} & 0 & A^{v_5}_{w^0} & 0 & 0 & 0 \\
0 & 0 & 0 & 0 & 0 & 0 & 0 & 0 & 0 & 0 & 0 & 0 & 0 & 1 & 0 & 0 \\
0 & 0 & 0 & 0 & 0 & 0 & 0 & 0 & 0 & 0 & 0 & 0 & 0 & 0 & 1 & 0 \\
0 & 0 & 0 & 0 & 0 & 0 & 0 & 0 & 0 & 0 & 0 & 0 & 0 & 0 & 0 & 1 \\
0 & A^{w}_{u_1^1} & A^{w}_{v_1^0} & 0 & 0 & A^{w}_{u_3^1} & A^{w}_{v_3^0} & 0 & 0 & A^{w}_{u_5^1} & A^{w}_{v_5^0} & 0 & A^{w}_{w^0} & 0 & A^{w}_{w^2} & 0
\end{bmatrix}$$

$$\tilde{f}_n(x,s) = \{0 \; 0 \; 0 \; 0 \; 0 \; 0 \; 0 \; 0 \; 0 \; 0 \; 0 \; 0 \; 0 \; 0 \; 0 \; \tilde{f}^{w}\}^T$$

References

[1] Kerwin E M 1959 *J. Acoust. Soc. Am.* **31** 952-62
[2] Mead D J and Markus S 1969 *AIAA Journal.* **10** 163-75
[3] Yeh J Y and Chen L W 2006 *Compos. Struct.* **73** 53-60
[4] Zhang Q J and Sainsbury M G 2000 *Comput. Struct.* **74** 717-30
[5] Chen Y C and Huang S C 2002 *Int. J. Mech. Sci.* **44** 1801-21
[6] Ramesh T C and Ganesan N 1994 *J. Sound. Vib.* **171** 577-601
[7] Ramesh T C and Ganesan N 1994 *J. Sound. Vib.* **175** 535-55
[8] Ramesh T C and Ganesan N 1994 *J. Sound. Vib.* **175** 359-70

[9] Wang H J and Chen L W 2004 *Finite Elem. Anal. Des.* **40** 737-55
[10] Krishna B V and Ganesan N 2007 *J. Sound. Vib.* **303** 575-95
[11] Hu Y C and Huang S C 2000 *J. Comput. Struct.* **16** 577- 91
[12] Pan H H 1969 *J. Sound Vib.* **9** 338-48
[13] Cao X T, Zhang Z I and Hua H X 2011 *Appl. Math. Mech.* **32** 495-506
[14] Li E Q, Li D K, Tang G J and Lei Y J 2007 *Acta Aeronaut. Et Astronaut. Sin.* **28** 1487-93
[15] Xiang Y, Huang Y Y and Lu J 2008 *Appl. Math. Mech.* **29** 1587-600
[16] Saravanan C, Ganesan N and Ramamurti.V 2000 *Comput. Struct.* **75** 395-417
[17] Li E Q, Li D K and Tang G.J 2008 *Eng. Mech.* **25** 6-11

Vertical vibration analysis for elevator compensating sheave

Seiji Watanabe[1], Takeya Okawa[2], Daisuke Nakazawa[1] and Daiki Fukui[1]

[1] Advanced Technology R&D Center, Mitsubishi Electric Corp., Hyogo, Japan
[2] Inazawa Works, Mitsubishi Electric Corp., Aichi, Japan

E-mail: Watanabe.Seiji@ay.MitsubishiElectric.co.jp

Abstract. Most elevators applied to tall buildings include compensating ropes to satisfy the balanced rope tension between the car and the counter weight. The compensating ropes receive tension by the compensating sheave, which is installed at the bottom space of the elevator shaft. The compensating sheave is only suspended by the compensating ropes, therefore, the sheave can move vertically while the car is traveling. This paper shows the elevator dynamic model to evaluate the vertical motion of the compensating sheave. Especially, behavior in emergency cases, such as brake activation and buffer strike, was investigated to evaluate the maximum upward motion of the sheave. The simulation results were validated by experiments and the most influenced factor for the sheave vertical motion was clarified.

1. Introduction

Conventional elevators move vertically by the traction machine, which suspends the car and the counter weight by the rope. The rope weight which is installed in high-rise buildings becomes so heavy that the car and the counterweight have to hang the compensating rope, to keep the rope tension balance, at the traction machine. The compensating sheave applies the tension to the compensating rope, and the sheave is guided at the bottom of the elevator shaft. As the compensating sheave is suspended by the compensating rope, the sheave weight is equivalent to the rope tension. The applied tension prevents the rope from sway motion while the car is running.

Conventionally, a lot of research concerning elevator ride comfort, or evaluation of rope resonance condition, have been investigated. Some papers show the compensating rope vibration due to the building sway motion. However, there is not much research focusing on the dynamic behavior of the compensating sheave.

This paper shows the compensating sheave model which includes the car, counter weight and the compensating rope. The proposed model is verified by the experiment, and the model can clarify the mechanism of the sheave vertical motion. The simulation model can also show the influence of each elevator parameter against the sheave vertical motion.

2. Simulation model

Figure 1 shows the total elevator configuration, which includes the compensating sheave. The main rope and the compensating rope are the continuous body, and they are modeled by the divided mass and the connecting spring. However, the rope model in figure 1 is simply represented by a single spring, $k_{01} \sim k_{23}$.

The total kinetic energy and the potential energy are given by the following equations.

$$T = \frac{1}{2}J_0\dot{\theta}_0^2 + \frac{1}{2}J_3\dot{\theta}_3^2 + \sum_{i=1}^{3}\frac{1}{2}M_i\dot{x}_i^2 \tag{1}$$

$$\begin{aligned} U &= \frac{1}{2}k_{01}\left(r_0\theta_0 + x_1\right)^2 + \frac{1}{2}k_{02}\left(r_0\theta_0 - x_2\right)^2 \\ &\quad + \frac{1}{2}k_{13}\left(x_1 - x_3 + r_3\theta_3\right)^2 + \frac{1}{2}k_{23}\left(x_2 - x_3 - r_3\theta_3\right)^2 \end{aligned} \tag{2}$$

where, r_0 and r_3 are the rotation radii of the traction machine and the compensating sheave. The equation of motion is derived by the above equations, and it contains the three degrees of translational freedom $x_1 \sim x_3$, and the two degrees of rotational freedom θ_0, θ_3.

$$M\ddot{\boldsymbol{x}} + C\dot{\boldsymbol{x}} + K\boldsymbol{x} = F \tag{3}$$

$$\boldsymbol{x} = \left[x_1,\ x_2,\ x_3,\ \theta_0,\ \theta_3\right]^T \tag{4}$$

In equation (3), two equations of motion concerning the vertical translation x_3 and the rotation θ_3 are given by,

$$M_3\ddot{x}_3 + T_{13} + T_{23} = -M_3 g \tag{5}$$

$$J_3\ddot{\theta}_3 + r_3\left(-T_{13} + T_{23}\right) = 0 \tag{6}$$

where, T_{13} and T_{23} show each rope tension, and g is the gravitational acceleration.

$$T_{13} = k_{13}\left(x_3 - x_1 - r_3\theta_3\right)\ ,\quad T_{23} = k_{23}\left(x_3 - x_2 + r_3\theta_3\right) \tag{7}$$

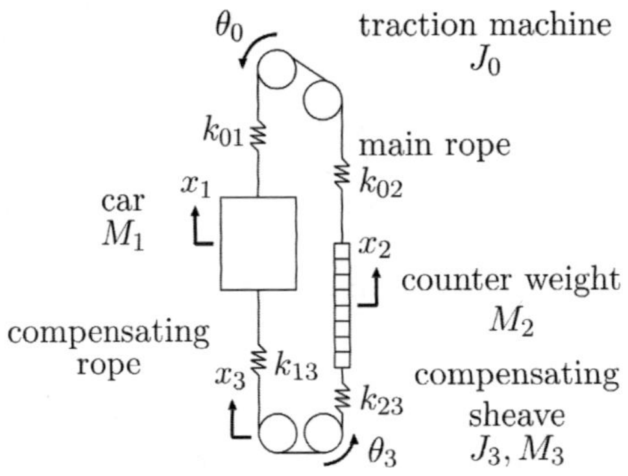

Figure 1. Configuration of elevator system.

3. Sheave vertical motion during normal operation

Based on the model in the previous section, the vertical motion of the compensating sheave is simulated, while the car is running at normal operation. The simulation results are compared with the experiments. Table 1 shows the simulation conditions.

Figure 2 shows the sheave vertical displacement, while the car moves downward from the highest floor to the lowest one. The zero position in vertical axis represents the initial displacement at the highest floor. The car accelerates from zero seconds to five seconds, and reaches the rated constant speed. Then the car decelerates from 45 seconds to 50 seconds.

The simulation result indicated by the broken line shows the same motion as the experiment in the solid line. Therefore, the proposed simulation model can evaluate the vertical motion of the compensating sheave properly.

Table 1. Simulation conditions.

car weight	9300 kg
counter weight	10600 kg
traveling distance	132 m
rated speed	180 m/min
car load	no load

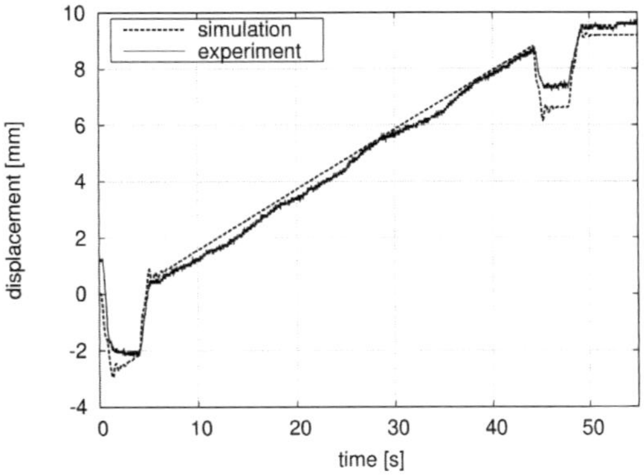

Figure 2. Sheave displacement during downward motion.

The following describes the mechanism of sheave vertical motion during the acceleration and deceleration. The car starts from the highest floor. Therefore, the stiffness of the car side main rope k_{01} in figure 1 is much higher than the counter weight side k_{02}. The car can trace the motion of the traction sheave, but the counter weight does not move in the same pattern as the traction sheave, due to the rope stiffness.

The simplified equation of motion for the main rope and the counter weight is derived by,

$$M_2\ddot{x}_2 + k_{02}\,(x_2 - r_0\theta_0) = -M_2 g \tag{8}$$

During acceleration, the counter weight receives upward acceleration, $\alpha > 0$. Therefore, the rope tension of the counter weight side is given by the following equation.

$$T_{02} = k_{02}\,(x_2 - r_0\theta_0) = -M_2(g + \alpha) \tag{9}$$

The above equation shows that the rope tension of the counter weight side increases during the acceleration. As the car position does not change, in spite of the rope stretch due to the increased tension of the counter weight side, the increased tension induces the vertical motion of the compensating sheave during the acceleration pattern from zero to five seconds.

In the same manner, the main rope tension of the car side during the deceleration near the lowest floor is given by the following equation.

$$T_{01} = k_{01}(x_1 + r_0\theta_0) = -M_1(g + \beta) \tag{10}$$

where, $\beta(> 0)$ is the upward deceleration. In equation (10), the rope tension increases during the deceleration, and the main rope of the car side stretches more than the counter weight side. The difference in stretch between both main ropes induces the downward motion of the compensating sheave during the deceleration motion from 45 seconds to 50 seconds.

The next step is the evaluation of the compensating sheave motion during constant speed. Generally, the counter weight is heavier than the car, as shown in table 1. Therefore, the rope tension of the counter weight side is always larger than the car side.

The main rope stretch of the counter weight side at the highest floor is longer than the car side stretch at the lowest floor. When the zero position of the compensating sheave is defined at the highest floor, the compensating sheave moves upward gradually due to the stretch difference. This is the reason why the graph in figure 2 shows the right upward pattern.

When the car weight is equivalent to the counter weight, the graph in figure 2 shows a flat pattern during constant speed conditions. On the other hand, the graph shows right downward pattern, if the car weight is heavier than the counter weight due to the number of rated passengers.

Figure 3 shows the upward motion from the lowest floor to the highest floor. The result in figure 3 is the same pattern as figure 2 by flipping the lateral axis. From the result, we can conclude that the running direction has no effect on the sheave vertical motion.

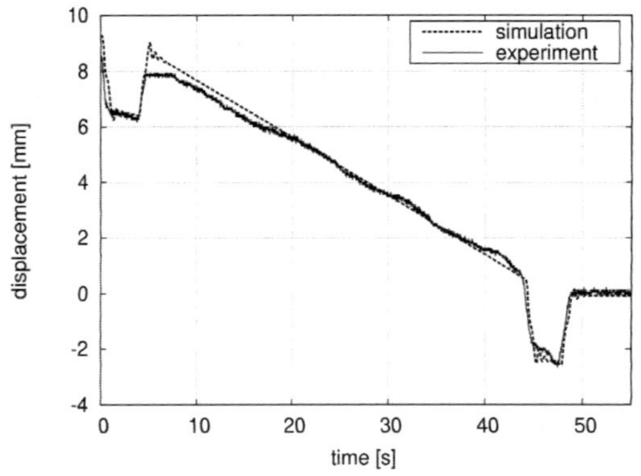

Figure 3. Sheave displacement during upward motion.

4. Sheave vertical motion during emergency brake operation

The compensating sheave has a limit switch which prevents the sheave from an excessive upward motion. The switch works during the safety gear operation or the buffer strike, and it should not be applied during the emergency brake operation. Therefore, it is important to evaluate the maximum upward displacement while using the emergency brake, to fulfill the design requirement.

The car deceleration while the emergency brake is engaged, changes according to the car running direction and the car load condition. As the car deceleration is higher, the vertical motion of the compensating sheave increases more. Therefore, the following conditions are investigated because of the larger deceleration occurrence.

(i) downward motion without car loading
(ii) upward motion with the rated car loading

As described in the previous section, the sheave vertical motion shows the right downward pattern when the car has a rated load. It is important to check the upward motion of the sheave, the proper simulation condition is the condition of (i).

Figure 4 shows the sheave vertical motion while the emergency brake is in operation near the highest floor. Please note that the acceleration in the following simulation is higher than in the previous section.

At the highest floor, the sheave motion is affected by the main rope stretch of the counter weight side. Therefore, equation (9) is used for evaluation. In equation (9), the acceleration $\alpha > 0$ is assigned to the brake deceleration $\gamma > 0$. The rope tension decreases as follows,

$$|T_{02}| = M_2(g - \gamma) < M_2 g \tag{11}$$

By the above equation, the compensating sheave moves upward during the emergency brake operation. In figure 4, the car accelerates from zero to four seconds, and then the emergency brake works past five seconds. The sheave motion is exactly the same as described in the above sentences. During the emergency brake operation from six seconds to eight seconds, the compensating sheave keeps the same position. When the car stops finally by the brake operation, the sheave receives a large vertical motion. The sheave shows a residual vibration, and the frequency matches the natural frequency due to the counter weight and the main rope.

Even though the peak residual vibration is larger than the experiment, the overall vibration pattern is the same as the experiment. From figure 4, it is important to check the timing of the emergency brake activation and the car stopping time.

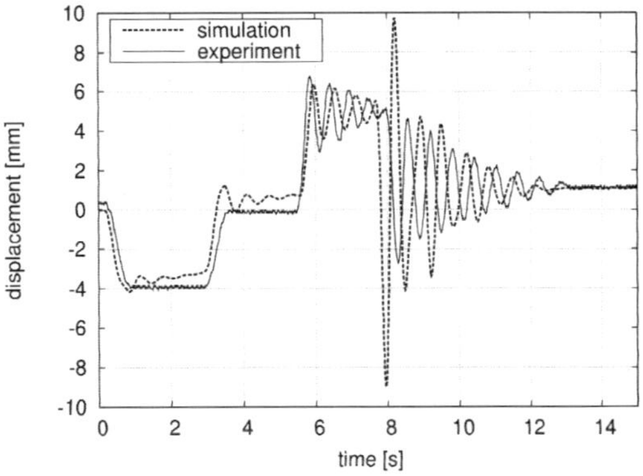

Figure 4. Brake stop near the top floor.

Figure 5 shows the emergency brake operation during the downward motion near the lowest floor. Figure 5 is the magnified scale near the time range of emergency brake. The simulation result matches the experiment.

As the sheave motion at the lowest floor is affected by the main rope of the car side. Equation (10) is used to evaluate the tension change. When the deceleration due to the emergency brake γ is applied upward to the car, the rope tension is given by,

$$|T_{01}| = M_1(g + \gamma) > M_1 g \tag{12}$$

When the emergency brake works at 43 seconds, the compensating sheave moves downward as shown in figure 5. The car also receives a large vibration at the instant of the car stop time at 45 seconds. The above large vibration is caused by the natural frequency of the car and the main rope.

As the sheave moves downward during the emergency brake operation, we can only focus on the residual vibration at the car stopping time.

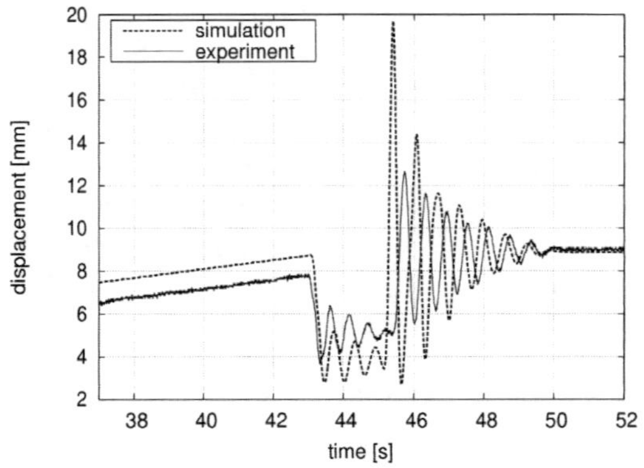

Figure 5. Brake stop near the bottom floor.

Concerning the emergency brake operation, we can say that the sheave moves upward near the highest floor in figure 4, and it moves downward near the lowest floor in figure 5. From these results, we can estimate that the sheave does not move vertically near the middle floor. To evaluate the assumption, figure 6 shows the simulation result during the emergency brake operation near the middle floor.

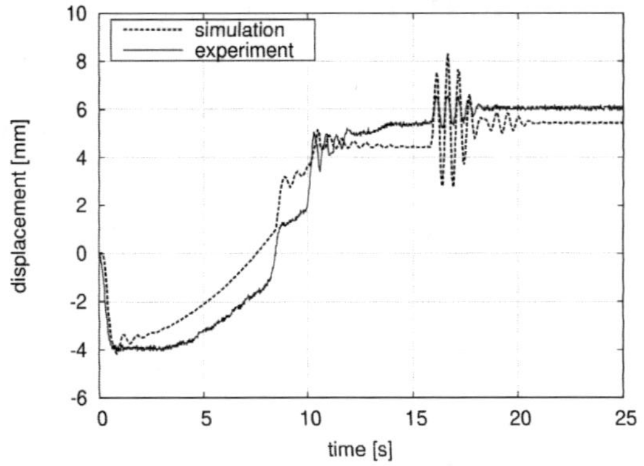

Figure 6. Brake stop at the middle floor.

In figure 6, the car speed increases to 540 m/min, and the car continues accelerating for the first eight seconds. After that, the car reaches the middle floor and the emergency brake begins working at 10 seconds. The sheave does not move vertically in the simulation or the experiment between 10 and 16 seconds while the brake is in operation.

From the above results, the highest upward motion occurs when the car moves downward with no car loading, and emergency brakes are working, near the lowest floors. The peak value appears at the instant when the car stops due to the brake activation.

In the following, each elevator parameter is checked to evaluate the relation with the vertical motion of the compensating sheave. As described before, the worst condition occurs near the lowest floors, but the following simulations are evaluated near the highest floors. This is because the relative evaluation is enough to check the parameter sensitivity.

We can guess that the stiffness of the main rope affects the sheave motion. To evaluate the stiffness change, several conditions of different building heights are evaluated and compared with each other. According to the building height, the main rope stiffness changes and the result is shown in figure 7.

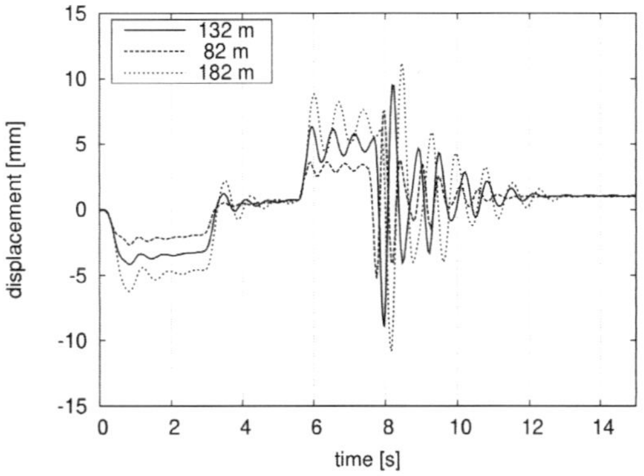

Figure 7. Effect of traveling distance.

In figure 7, the building height of 132 m in solid line is the original pattern, and the other conditions are adjusted to the zero position at the start time. In the case of 82 m shown in the broken line, the rope stiffness increases due to the short rope length. Therefore, the sheave vertical motion is less than the original height. On the other hand, in the case of 182 m shown in the dotted line, the rope stiffness decreases and the sheave shows larger vibration.

From the above result, we can calculate the lowest rope stiffness necessary in the most severe cases, as in the rope parameter study.

In the next step, the sheave weight influence is evaluated. Figure 8 shows the weight variation results by 50% increase and decrease.

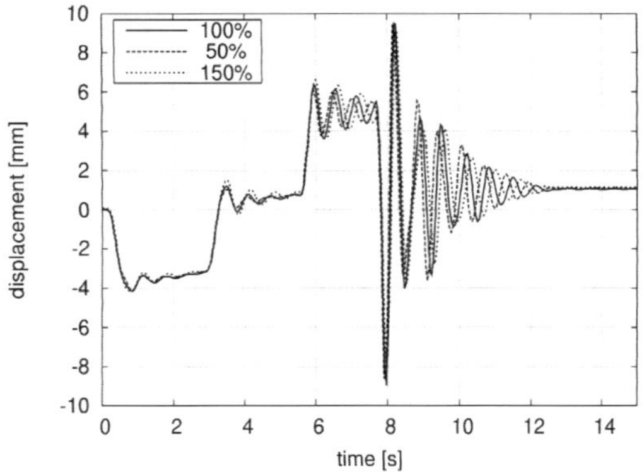

Figure 8. Effect of compensating sheave weight.

As shown in figure 8, the weight change does not affect the sheave's vertical motion. Therefore, the sheave motion is mostly governed by both ends of the compensating rope.

5. Sheave vertical motion during buffer strike

In the previous section, the emergency brake operation is evaluated for the sheave vertical motion. In this section, the buffer strike operation is investigated.

When the car does not stop at the lowest floor, the car can stop any further downward motion by the buffer, which is installed at the bottom of the elevator shaft. Figure 9 shows the sheave's vertical motion during the buffer strike. The simulation conditions are given in table 2.

Table 2. Simulation conditions.

car weight	4600 kg
counter weight	5400 kg
traveling distance	147 m
rated speed	210 m/min
car load	no load

In figure 9, the car passes the lowest floor at the start time, and the car strikes the buffer. The sheave weight is decreased and increased by 50%. From the simulation results, we can see that the sheave weight variation does not affect the sheave's vertical motion.

Figure 10 shows the corresponding car and counter weight motion. When the car strikes the buffer as shown in the solid line, the car stops at 0.5 seconds. On the other hand, the counter weight continues to move upward by its inertia force. Therefore, the compensating sheave also follows the upward motion of the counter weight during the first 0.7 seconds. After that, the counter weight begins falling by gravitational force, and the compensating sheave also moves downward until 1.4 seconds. When the rope tension of the counter weight side recovers due to falling motion, the car and the counter weight cause vibration around 1.5 seconds. At that time, the sheave receives two vibrations; one is induced by the car side motion of 0.9 second time period, and the other is 0.4 second time period due to the counter weight vibration.

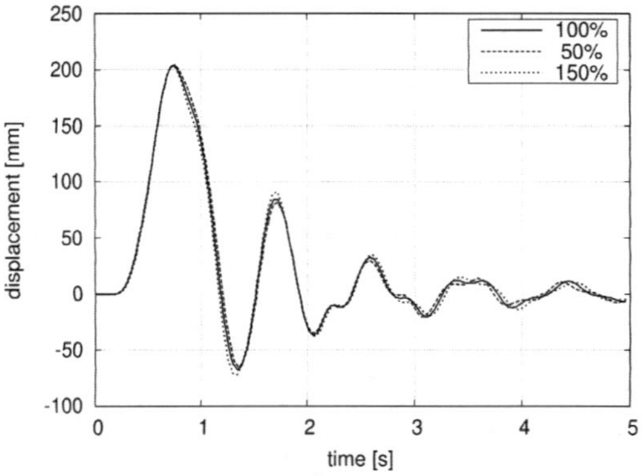

Figure 9. Effect of compensating sheave weight.

From the above simulation results, we can say that the sheave motion is governed by the car and counter weight motion, and the sheave weight does not affect the upward motion.

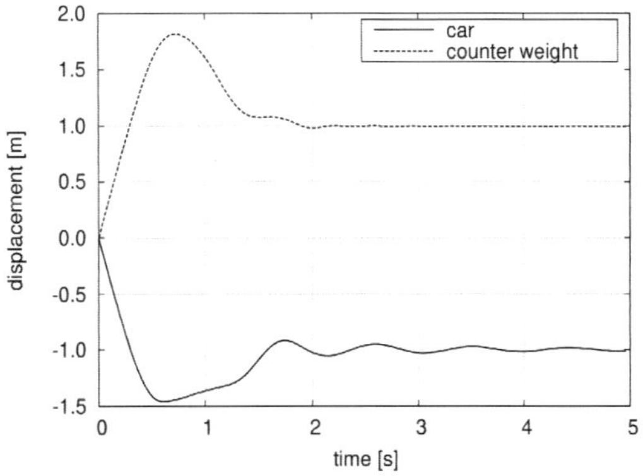

Figure 10. Transient motion of car and counter weight.

6. Conclusions

This paper showed the elevator dynamic model to calculate the compensating sheave motion. The derived simulation model can show the same vertical behavior as the experiment results. In the case of normal traveling conditions, we can reach the conclusion that the weight difference between the car and the counter weight affects the transient motion of the sheave vertical direction while the car is traveling.

We can also reach the following conclusions;

(1) The maximum upward motion during emergency brake operation, occurs in the no-loaded condition with the downward traveling near the lowest floors.

(2) Lower rope stiffness of the main rope induces larger upward motion of the compensating sheave.

(3) The weight of the compensating sheave has no influence on its upward motion.

Journal of Physics: Conference Series **448** (2013) 012008 doi:10.1088/1742-6596/448/1/012008

Application of Higher-Order Cumulant in Fault Diagnosis of Rolling Bearing

Yongjun Shen, Shaopu Yang, Junfeng Wang

Department of Mechanical Engineering, Shijiazhuang Railway Institute, Shijiazhuang 050043, China

E-mail: shenyongjun@126.com
E-mail: yangsp@sjzri.edu.cn
E-mail: 450111043@qq.com

Abstract. In this paper a new method of pattern recognition based on higher-order cumulant and envelope analysis is presented. The core of this new method is to construct analytical signals from the given signals and obtain the envelope signals firstly, then compute and compare the higher-order cumulants of the envelope signals. The higher-order cumulants could be used as a characteristic quantity to distinguish these given signals. As an example, this method is applied in fault diagnosis for 197726 rolling bearing of freight locomotive. The comparisons of the second-order, third-order and fourth-order cumulants of the envelope signals from different vibration signals of rolling bearing show this new method could discriminate the normal and two fault signals distinctly.

1. Introduction

In recent years, Higher-Order Statistics (HOS) has received increasing interest in signal processing [1]. It has also been used in many other fields, such as communication, sonar, biomedicine, speech signal, image processing, et al [2-5]. As we all know, there are some shortcomings of second-order statistics used in signal processing, such as inability to analyze the phase relation between different harmonic signals, and invalidity for nonlinear and non-Gaussian signals. On the other hand, application of HOS in signal processing could avoid these problems, due to its abilities including suppressing additive colored noise, identifying non-minimum phase systems, extracting information due to deviations from Gaussianity, detecting and characterizing nonlinear properties in signals or systems, analyzing the cyclostationary time series, separating unknown sources or Blind Source Separation, et al.

In this paper a new method of pattern recognition is presented, which is based on the envelope analysis of the given signals and the corresponding higher-order cumulant, one kind of HOS. The procedure of this method is to construct analytical signals by Hilbert Transform from the given signals, and obtain the envelope signals at first, then compute and compare the higher-order cumulants of these envelope signals. The higher-order cumulants could be used as the characteristic quantity to discriminate these given signals. In order to illustrate the validity of this method, it is used to discriminate the measured signals from 197726 rolling bearings of freight locomotive in different conditions. The comparison of the second-order, third-order and fourth-order cumulants of different vibration signals of rolling bearing show this new method could discriminate the normal and two fault signals distinctly.

2. Definition and Properties of Higher-Order Cumulant

In this part we would review the definition and properties of HOS curtly, which is the basis of the new method of pattern recognition.

2.1. Definition of Higher-Order Cumulant

Letting x denote a random variable and $f(x)$ its probability density function, its moment generating function could be defined as

$$\phi_x(\omega) = \mathrm{E}[\exp(j\omega x)] = \int_{-\infty}^{\infty} f(x)\mathrm{e}^{\mathrm{j}\omega x}\,\mathrm{d}x, \tag{1}$$

where E denotes the expectation operator.

The definition of cumulant generating function of x is

$$C(\omega) = \ln\{\mathrm{E}[\exp(j\omega x)]\}. \tag{2}$$

Considering the random vector $\mathbf{x} = [x_1, x_2, \cdots, x_n]^{\mathrm{T}}$ composed of n random variables, and supposing $\boldsymbol{\omega} = [\omega_1, \omega_2, ..., \omega_n]^{\mathrm{T}}$, one could obtain the definition of joint characteristic function for $\mathbf{x}$ as follow

$$\phi_{\mathbf{x}}(\omega_1, \omega_2, \cdots, \omega_n) = \mathrm{E}[\exp\mathrm{j}(\omega_1 x_1 + \omega_2 x_2 + \cdots + \omega_n x_n)], \tag{3}$$

and it could also be written as

$$\phi_{\mathbf{x}}(\boldsymbol{\omega}) = \mathrm{E}[\exp(j\boldsymbol{\omega}^{\mathrm{T}}\mathbf{x})]. \tag{4}$$

And the cumulant generating function of $\mathbf{x}$ is defined as

$$C(\boldsymbol{\omega}) = \ln\{\mathrm{E}[\exp\mathrm{j}(\boldsymbol{\omega}^{\mathrm{T}}\mathbf{x})]\}. \tag{5}$$

Then the joint cumulant of $\mathbf{x}$ is

$$\mathrm{cum}(\mathbf{x}) = (-\mathrm{j})^k \frac{\partial^k \ln \phi_{\mathbf{x}}(\boldsymbol{\omega})}{\partial\omega_1^{k_1}\partial\omega_2^{k_2}\cdots\partial\omega_n^{k_n}}\Big|_{\omega_1=\omega_2=\cdots\omega_n=0}, \tag{6}$$

where $k = (k_1 + k_2 + \cdots + k_n)$.

Supposing $\{x(t)\}$ as k-order stationary random process and meet the condition of ergodicity, the k-order cumulant of $\{x(t)\}$ could be defined as the k-order joint cumulant of $[x(t), x(t+\tau_1), \cdots, x(t+\tau_{k-1})]^{\mathrm{T}}$, that is to say,

$$C_{k,x}(\tau_1, \tau_2, \cdots, \tau_{k-1}) = \mathrm{cum}[x(t), x(t+\tau_1), \cdots, x(t+\tau_{k-1})], \tag{7}$$

where $k_1 = k_2 = \cdots k_n = 1$.

According to the hypothesis of stationarity, it could be concluded that the k-order cumulant of $\{x(t)\}$ is only the function of $k-1$ time intervals $\tau_1, \tau_2, \cdots, \tau_{k-1}$. Then one could obtain the equations

$$C_{2,x}(\tau_1) = \mathrm{E}[x(t)x(t+\tau_1)], \tag{8}$$

$$C_{3,x}(\tau_1, \tau_2) = \mathrm{E}[x(t)x(t+\tau_1)x(t+\tau_2)], \tag{9}$$

$$C_{4,x}(\tau_1, \tau_2, \tau_3) = \mathrm{E}[x(t)x(t+\tau_1)x(t+\tau_2)x(t+\tau_3)]$$

$$-C_{2,x}(\tau_1)C_{2,x}(\tau_2-\tau_3)-C_{2,x}(\tau_2)C_{2,x}(\tau_3-\tau_1)$$

$$-C_{2,x}(\tau_3)C_{2,x}(\tau_1-\tau_2). \tag{10}$$

2.2. Common Properties of Higher-order Cumulant

(1) The higher-order cumulant meets the symmetry between all the random variables, i.e.

$$\text{cum}(x_1,x_2,\cdots,x_k)=\text{cum}(x_i,x_j,\cdots,x_q) \tag{11}$$

where $(i, j, \cdots, q)$ is an arbitrary permutation form of $(1, 2, \cdots, k)$.

(2) The higher-order cumulant is linear, namely,

$$\text{cum}(\lambda_1 x_1,\lambda_2 x_2,\cdots,\lambda_k x_k)=\text{cum}(x_1,x_2,\cdots,x_k)\prod_i \lambda_i, \tag{12}$$

where $\lambda_i, i=1, 2, \cdots, k$ are complex constants.

(3) The cumulant of the sum of two random variables equals the sum of the cumulant of the two variables, that is to say,

$$\text{cum}(x_0+y_0,z_1,z_2,\cdots,z_k)=\text{cum}(x_0,z_1,\cdots,z_k)+\text{cum}(y_0,z_1,\cdots,z_k), \tag{13}$$

where x_0 and y_0 are two different random variables. And this property could be extended to the case of many random variables.

(4) If x_i is mutually independent to y_i, one could obtain

$$\text{cum}(x_1+y_1,x_2+y_2,\cdots,x_k+y_k)=\text{cum}(x_1,x_2,\cdots,x_k)+\text{cum}(y_1,y_2,\cdots,y_k). \tag{14}$$

Accordingly, application of higher-order cumulant in signal processing would restrain the noise when non-Gaussian signal added by Gaussian noise.

(5) If λ is a constant, it has no effect on the result of cumulant, i.e.

$$\text{cum}(\lambda+x_1,x_2,\cdots,x_k)=\text{cum}(x_1,x_2,\cdots,x_k) \tag{15}$$

3. Analysis of the Envelope Signal

In this section we review the Hilbert Transform and the analysis of complex envelope spectrum.

3.1 Hilbert Transform

Hilbert Transform is linear, which maps a function to another one in the same space. It reveals the mutual relation between the real and the imagery part of the system function. Based on Hilbert Transform, some new communication and digital signal systems have been developed and used in many cases.

Letting $x(t)$ ($t\in(-\infty,+\infty)$) be a real function, its Hilbert Transform could be defined as

$$\hat{x}(t)=\int_{-\infty}^{+\infty}\frac{x(\tau)}{\pi(t-\tau)}\mathrm{d}\tau \tag{16}$$

which is also denoted by $\hat{x}(t)=\text{H}[x(t)]$. It could be found that $\hat{x}(t)$ is the convolution of $x(t)$ and πt, that means,

$$\hat{x}(t)=x(t)*\frac{1}{\pi t} \tag{17}$$

3.2 Analysis of complex envelope spectrum

Analysis of complex envelope spectrum is to recover the modulated signal from the complex analytical signal constructed by the measured signal based on Hilbert Transform. The procedure of this analysis could be illustrated as follow:

(1) obtaining the measured signal from the sensors which may be written as $x(t) = r(t)\cos(2\pi ft + \phi(t))$, where the amplitude modulation $r(t)$ and phase modulation $\phi(t)$ all exist.

(2) applying the Hilbert Transform to $x(t)$ so that one could obtain the $\hat{x}(t)$ which could approximately be expressed as $\hat{x}(t) = r(t)\sin(2\pi ft + \phi(t))$.

(3) constructing the analytical signal $z(t) = x(t) + j\hat{x}(t)$, where its real and imaginary part is $x(t)$ and $\hat{x}(t)$ respectively.

(4) extracting the modules of $z(t)$ as $|z(t)| = \sqrt{x^2(t) + \hat{x}^2(t)} = r(t)$ which is the amplitude modulated signal, and the instantaneous phase of $z(t)$ as $\theta(t) = \arctan\left(\frac{\hat{x}(t)}{x(t)}\right) = 2\pi ft + \phi(t)$ which is the original phase modulated signal after subtracting the ramp signal $2\pi ft$.

4. New method of pattern recognition based on higher-order cumulant and analysis of complex envelope spectrum

Based on the aforementioned description, one could establish the new method of pattern recognition based on higher-order cumulant and analysis of complex envelope spectrum. The first step of this method is to obtain the amplitude modulated signal from the measured signal. And the second step is to calculate the higher-order cumulant of and then compare them.

In order to illustrate the validity of this method, we apply it to recognize the measured signals from 197726 rolling bearings of freight locomotive. The results are shown in Fig 1 to Fig 3, where the line 1, 2, 3 denote the signal from normal bearing, rolling element defect bearing, and outer race defect bearing respectively. It could be found that, second-order and third-order cumulants of the measured signals could not distinguish the three operating conditions, whereas the fourth-order cumulants of the measured signals or the second-order, the third-order and the fourth-order cumulants of the envelope signals could distinguish these three conditions distinctly.

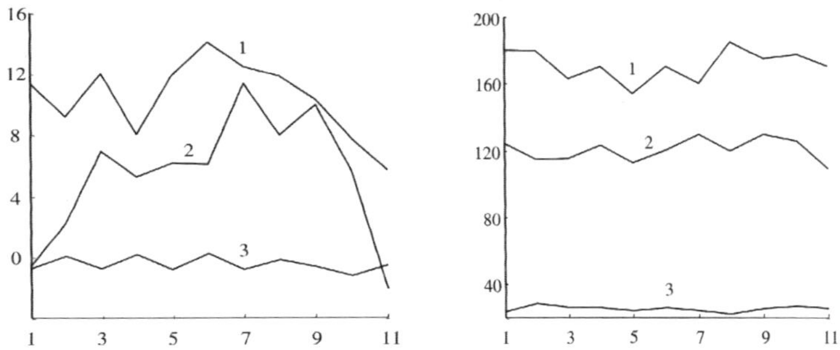

(a) (b)

Figure 1. The Second-order cumulant ((a) for measured signals and (b) for envelope signals)

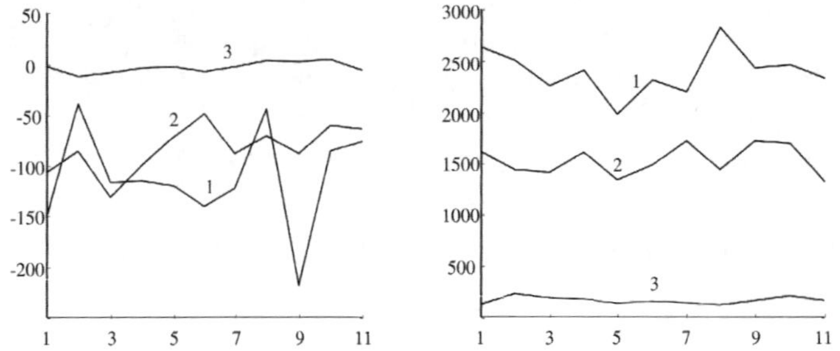

(a) (b)

Figure 2. The Third-order cumulant ((a) for measured signals and (b) for envelope signals)

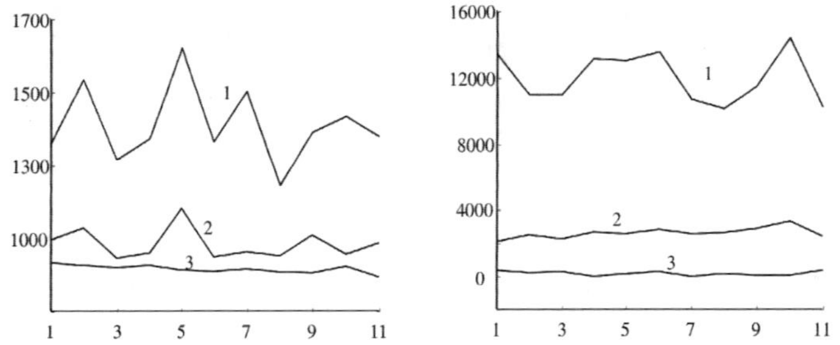

(a) (b)

Figure 3. The Fourth-order cumulant ((a) for measured signals and (b) for envelope signals)

5. Conclusions

In this paper a new method for pattern recognition is investigated, which is based on higher-order cumulant and envelope analysis. And the efficiency of this method is illustrated by applying it in recognition of the operating conditions of 197726 rolling bearings of freight locomotive, which could distinguish the three conditions of rolling bearings distinctly.

6. References

[1] Mendel J M 1991 Tutorial on Higher-Order Statistics (Spectra) in Signal Processing and System Theory: Theoretical Results and Some Applications *J.Proc. IEEE* **79** 277-305

[2] Zhang X D 1996 *Analysis of Time Series — Higher-Order Statistics* (Beijing: Qinghua University Press)

[3] Nikias and Petropulu A 1993 *Higher-Order Spectra: a Nonlinear Signal Processing Framework* (New Jersey: Prentice Hall)

[4] Tugnait J K 1997 Identification and deconvolution of multichannel linear non-Gaussian processes using higher order statistics and inverse filter criteria *J.Trans. Signal Processing IEEE* **45** 658-672

[5] Collis W B, White P R and Hammond J K 1998 Higher-Order Spectra: The Bispectrum and Trispectrum *J.Mechanical System and Signal Processing* **12** 375-394

Acknowledgments

The authors are grateful to the support by National Natural Science Foundation of China (No. 11072158 and 10932006), Natural Science Funds for Distinguished Young Scholar of Hebei Province (No. E2010002047), the Program for New Century Excellent Talents in University (NCET-11-0936), and the Program for Changjiang Scholars and Innovative Research Team in University (No. IRT0971).

Critical load and buckling of the single pile foundation subjected to the vertical load

Jianjun Ma[1], **Jian Peng**[2], **Lianhua Wang**[1,3] **and Yueyu Zhao**[1]

[1] College of Civil Engineering, Hunan University, Changsha, Hunan 410082, PR China
[2] College of Mechanical and Vehicle Engineering, Hunan University, Changsha, Hunan 410082, PR China.
[3] Key Laboratory of Building Safety and Energy Efficiency, Ministry of Education, Hunan 410082, PR China

E-mail: majianjun@hnu.edu.cn

Abstract. In this study, the critical load and buckling of the single pile foundation subjected to the vertical load are investigated. Considering the second-order moment of the soil-structure interaction, the refined model of the single pile foundation is derived. Then, the critical load and buckling phenomenon of the single pile foundation is examined. Moreover, the effects of the vertical load and the foundation parameters on the critical load and buckling of the single pile foundation are systematically investigated.

1. Introduction

In the civil engineering, the soil-structure interaction problem is an important part of the dynamic design and analyze of the structure. In general, the buckling of the substructure plays an important role in the seismic behavior of the civil engineering. Based on the engineering practice, the foundation of the structure usually can define as the slender structure. Obviously, the pile foundation is a typical vertical slender structure.

The pile foundation is the usual deep foundation, which can bear a variety of the static and dynamic load. As we know, numerous studies have addressed the theory research of the pile foundation [1, 2].

To meet the needs of the design of the pile foundation, some researches of the dynamic stability and buckling of the pile foundation have been published. Gabr and his co-workers proposed the computation model of the critical load of the slender friction pile foundation [3, 4]. Based on the engineering facts, Kuo et al studied the minimum embedded length of the monopiles subjected to the cyclic horizontal load [5]. However, there only few studies have focused on the buckling of the pile foundation with the second-order moment of the soil-structure interaction.

In this paper, considering the second-order moment of soil-structure interaction, the refined model of the pile foundation is proposed. Then, the critical load and buckling of the pile are investigated. Finally, numerical analysis of the effect of the main parameters on the buckling phenomena of the pile foundation is examined.

2. Theoretical procedure

2.1. Geometry and Loading of the pile

In this study, the attention is focused on the slender circular pile foundation with length H, diameter 2R, and subjected to the lateral load $p(t)$ ($p(t) = P\cos\Omega t$, where P and Ω denote the amplitude and frequency of the excitation) and the vertical load $F(t)$. As show in Fig. 1, a cylindrical coordinate system $O-r\varphi z$ is chosen, with the origin O placed at the upper end of the undeformed pile and the z direction along the axes of pile. In this paper, the subgrade reaction of the elastic foundation denoted as $q(z,t) = k_0 v(z,t) - k_1 v''(z,t)$, where k_0 and k_1 are the Winkler parameter and the shear parameter of the soil medium, respectively [6]. Obviously, the displacements of the pile and the soil medium around the pile can be denoted as the separable functions of the cylindrical coordinate r, φ and z. Accordingly, the displacements of the pile can be assumed as $v(z,t)$ and $w(z,t)$, which are the displacement along the pile axis and the vertical displacement of the pile, respectively.

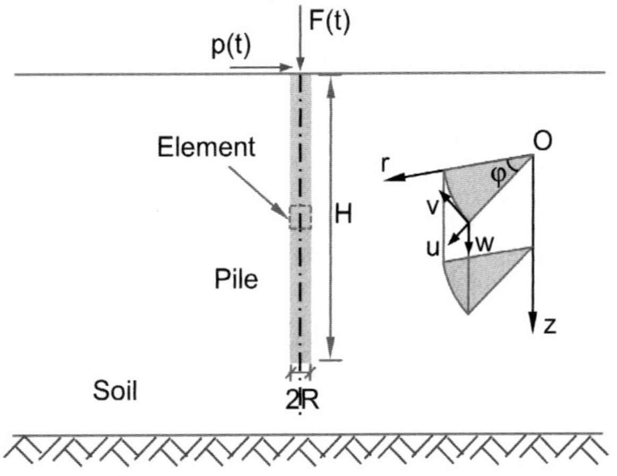

Figure 1. Pile-soil interaction system.

2.2. Strain and stress of the pile

Referring to Fig. 1, the axial elongation of the pile element can be written as [7]

$$e = \sqrt{(1+w')^2 + v'^2} - 1 \tag{1}$$

where the prime indicates differentiation with respect to z.

Moreover, based on the displacement vector of the pile in the local coordinate system, the nonzero axial strain of the pile can be obtained

$$\varepsilon_z = e - r\cos(\varphi)\kappa \tag{2}$$

where $\kappa = \theta'$ is the curvature, θ is the rotation angle.

2.3. Equation of motion

Considering the compressive deformation of the pile, the strain energy U of the pile can be written as

$$U = \int_0^H \frac{1}{2} EI\kappa^2 \mathrm{d}z + \int_0^H \frac{1}{2} EAe\mathrm{d}z \tag{3}$$

where E is the Young's modulus of the pile; I is the moment of inertia of the cross-section; A is the area of the cross-section.

The kinetic energy T is:

$$T = \int_0^H \frac{1}{2}\big[\rho A(\dot{w}^2 + \dot{v}^2) + j\dot{\theta}^2\big]\mathrm{d}z \tag{4}$$

where ρ is the density of the pile; j is the rotary inertia.

Applying the Hamilton principle, the motion equation of the pile foundation can be obtained

$$\rho A\ddot{v}+c\dot{v}+EIv''''-j\ddot{v}''+k_0v-k_1v''+EAF(t)v'' = p\delta(z)-R\big[v'(k_0v-k_1v'')\big]'+\frac{EA}{2H}v''\int_0^H v'^2\mathrm{d}z \tag{5}$$

where the damping term has been added with the viscous damping coefficient c; $\delta(z)$ is the Dirac δ function.

To obtain a more general conclusion of the dynamic response of the pile foundation, the following nondimensional variables are introduced

$$\begin{gathered} z^* = \frac{z}{H};\quad v^* = \frac{v}{H};\quad \lambda = \frac{R}{H};\quad J = \frac{j}{\rho AH^2};\quad t^* = t\sqrt{\frac{EI}{\rho AH^4}};\quad F^* = \frac{FH^2}{r^2} \\ c^* = \frac{cH^2}{\sqrt{\rho AEI}};\quad K_0 = \frac{k_0H^4}{EI};\quad K_1 = \frac{k_1H^2}{EI};\quad P^* = \frac{PH^4}{EI};\quad \Omega^* = \Omega\sqrt{\frac{\rho AH^4}{EI}}. \end{gathered} \tag{6}$$

where $r = \sqrt{I/A}$ is the radius of gyration. Substituting Eq. (6) into Eq. (5), we can obtain

$$\ddot{v} + c\dot{v} + Lv = P\delta(z)\cos\Omega t + G_2(v,v) + \frac{2v''}{\lambda^2}\int_0^1 v'^2\mathrm{d}z, \tag{7}$$

where the terms of rotary inertia has been neglected; the asterisks are dropped for simplicity; L is the linear operator, $Lv = v'''' + Fv'' - K_1v'' + K_0v$; G_2 is the quadratic nonlinear operator, $G_2(v,v) = -\lambda(K_0vv' - K_1v'v'')'$. Accordingly, the boundary conditions can be written as

$$z = 0,\ 1:\quad v(z,t)v''(z,t) = 0;\quad v(z,t)v'''(z,t) = 0;\quad v'(z,t)v''(z,t) = 0. \tag{8}$$

2.4. Critical load and buckling

Referring to Eq. (7), the critical load exists which can be obtained relevant to the buckling of the pile, when the pile subjected to the lateral and vertical loads. If all the damping and lateral load terms, the term related to the variable t, and the second-order moment term are neglected, the motion equation of the pile can be written as

$$\psi'''' + (F - K_1)\psi'' - \frac{2\psi''}{\lambda^2}\int_0^1 \psi'^2\mathrm{d}z + K_0\psi = 0, \tag{9}$$

where $\psi(z)$ is the buckling shape of the pile. If the $\psi(z)$ is a known function, the following parameter can be included

$$\Gamma = \frac{2}{\lambda^2}\int_0^1 \psi'^2\mathrm{d}z, \tag{10}$$

Substituting Eq. (10) into Eq. (9), we can obtained

$$\psi'''' + T^2\psi'' + K_0\psi = 0, \tag{11}$$

where $T^2 = F - K_1 - \Gamma$ is the constant value, which can be used to determine the critical load.

3. Results and discussions

Based on the characteristic equation, Eq. (11), we investigate the buckling phenomenon and frequency of the pile foundation. In this numerical analysis, the dimensional parameters and material properties of the pile and the soil medium are as follows: $E_s = 23.94$MPa, $\nu_s = 0.2$, $\nu = 0.25$, $H = 6.096$m, $R = 0.243$m, $\rho = 2.403 \times 10^3$kg/m^3, $E = 2.482 \times 10^4$MPa, where E_s is the Young's modulus of the soil medium; ν_s is the Poisson ratio of the soil medium; ν is the Poisson ratio of the pile. In the following numerical analysis, we consider two cases of the boundary conditions: free-hinged and free-free. Moreover, the effect of the boundary constraint on the buckling of the pile also be investigated.

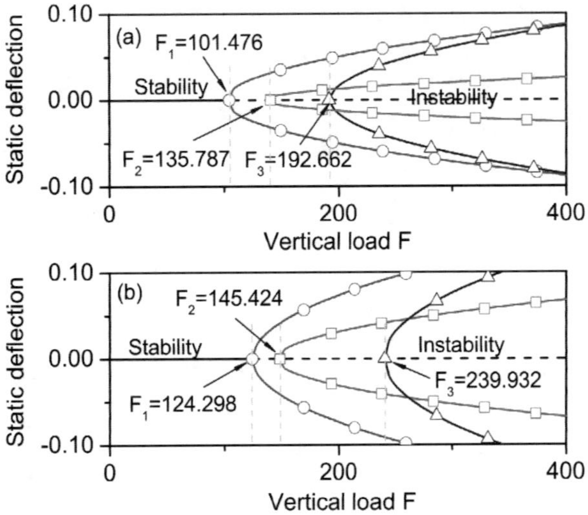

Figure 2. Bifurcation of the static deflection and the critical load of the pile: a) free-hinged; b) free-free.

Fig. 2 shows the bifurcation phenomena and the critical load of the pile. In general, the pile foundation not exhibits buckling and the static deflection (stability), until the vertical load greater than the critical load F_{cr}. Referring to Fig. 2, when the vertical load beyond the second-order critical load $F_2 = 135.787(145.424)$, there are three cases of the equilibrium state, which including the line mode (instability) and the buckling modes of the first- or second-order critical load. Moreover, compared to the free-hinged pile foundation, the static deflection of the free-free pile foundation quickly increases with the increase of vertical load.

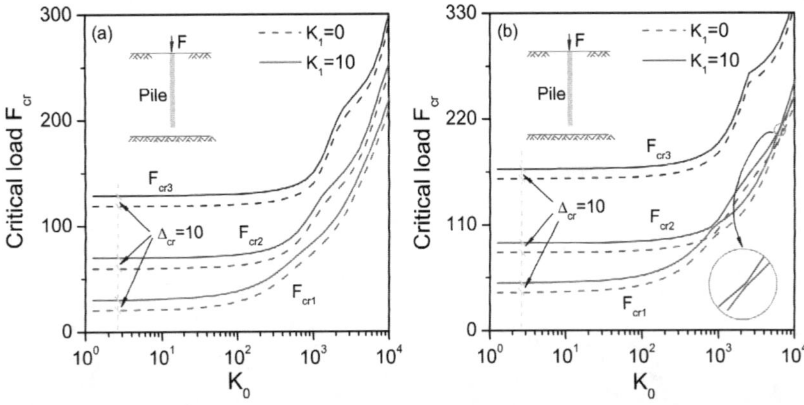

Figure 3. Effect of the foundation parameters on the critical load: a) free-hinged; b) free-free.

To reveal the effect of the foundation parameters on the critical load F_{cr}, Fig. 3 shows the variation of the first three order critical loads with the Winkler parameter of the foundation. Obviously, the critical load increases with the increase of the shear parameter K_1, and the increase value equal to K_1 ($F_{cr} = K_1 + T^2$). In general, the relaxation of the boundary constraint causes the increase of the critical load. As shown in Fig. 3, the growth trend of the critical load is very weak when the Winkler parameter $K_0 < 100$. However, when the Winkler parameter increases beyond the specific value, the effect of the Winkler parameter on the critical load is very significant. On the other hand, it is interesting that the first and second-order critical load is very closer to each other, when the pile foundation with the free-free constraint and $K_0 = 794(6310)$.

4. Conclusions

An analytical model of the pile foundation subjected to the vertical load considering the second-order moment of the soil-structure interaction is proposed in this study. Then, the critical load and buckling of the pile are investigated. Numerical results shown that the effect of the vertical load on the buckling is significant, the boundary conditions and the foundation parameters are also the important factors in the analysis of the buckling of the pile foundation. Moreover, the quadratic nonlinear term included in the motion equation of the pile foundation, we would better to research the nonlinear response of the pile foundation in the following study.

Acknowledgments

The authors are grateful to the National Natural Science Foundation of China (NNSFC) under Grant No. 10972073 and 11032004 for financial support in this study.

References

[1] Poulos H C 1989 Pile behaviour: theory and application *Géotechnuque* **39** 365-415.
[2] Li R and Gong J 2009 Analysis of laterally loaded pile in layered soils *Electronic Journal of Géotechnical Engineering* **13** J: 1-16.
[3] Gabr M A, Wang J and Kiger S A 1994 Effect of boundary conditions on buckling of friction piles *Journal of Engineering Mechanics* **120** 1392-1400.
[4] Gabr M A, Wang J and Zhao M 1997 Buckling of piles with general power distribution of lateral subgrade reaction *Journal of Géotechnical and Géoenvironmental Engineering* **123** 123-130.
[5] Kuo Y S, Achmus M and Abdel-Rahman K 2012 Minimum embedded length of cyclic horizontal load monopiles *Journal of Géotechnical and Géoenvironmental Engineering* **138** 357-363.
[6] Wang L H, Ma J J, Zhao Y Y and Liu Q J 2012 Refined modelling and free vibration of inextensional beams on elastic foundation *Journal of Applied Mechanics* DOI: 10.1115/1.4023032.
[7] Nayfeh A H and Pai P F 2004 *Linear and nonlinear structural mechanics* (New York: Wiley-Interscience)

Journal of Physics: Conference Series **448** (2013) 012010 doi:10.1088/1742-6596/448/1/012010

Determination of the response distributions of cantilever beam under sinusoidal base excitation

Wei Sun, Ying Liu, Hui Li and Dejun Pan
School of Mechanical Engineering & Automation
Northeastern University, Shenyang 110819, China

E-mail: weisun@me.neu.edu.cn

Abstract. As a kind of base excitation, shaking table is often used to test the dynamic characteristics of structures. However, the prediction of response to base excitation hasn't been solved effectively, which limits the further research on the test and analysis method with respect to base movement. This article is based on a cantilever beam and focuses on its response prediction under sinusoidal base excitation. By moment and force equilibrium equations, an analytical model is built for this cantilever beam, and then a method to predict dynamic response at base excitation is proposed. Finally, the method is used to solve the vibration response distributions of the cantilever beam at base excitation. Correctness of this method is also proved by comparing the result with experimental data.

1. Introduction

Base excitation is an important way to test vibration parameters of a structure. It is widely used because of stable input energy and being close to real constrained condition [1-4]. Yoshikazu Araki[1] tested the response of an object placed on a vibration isolator by shaker. Francesco Pellicano[2] tested the nonlinear parameters of cylindrical shell by shaker with a sine excitation.

However, the analysis of structure under base excitation progressed slowly compared with the experiment mentioned above. Most scholars generally adopted FEM to analyze the dynamic characteristics. For example, Liu[5] offered a method that simplified base excitation to distributed force by means of equivalent loads and create a FEM model to solve the problem. Allahdadian[6] proposed a topology optimization scheme which was used to reduce the vibration response under the base excitation by FEM.

Some scholars[7-9] also were devoted to doing theory analysis for the structure under base excitation. For example, on the basis of neglecting the mass of the beam, a composite structure at the base exciting was analyzed by Feng[7]. Kocaturk[8] analyzed the response of a viscoelastic support

Journal of Physics: Conference Series **448** (2013) 012010 doi:10.1088/1742-6596/448/1/012010

cantilever beam under sinusoidal base excitation by analytical method, but ignored the damping of the beam itself and only considered the damping and stiffness of the support. The analysis results will be inaccurate if the vibration characteristics of the structure at base excitation are not fully introduced.

Therefore, a detailed analytical analysis of the vibration response of a cantilever beam under the base excitation was done in this paper. Firstly, the dynamic differential equation of cantilever beam system was established considered the base excitation and the system damping. Secondly, modal superposition method was used to solve the equation and the response distributions of cantilever beam under sinusoidal base excitation were acquired. Finally, the analysis method was demonstrated using a study case and the validity of the proposed method was verified by the comparison of the analytical solution and the experimental results.

2. Analytical model of cantilever beam under base excitation

The cantilever beam system under base excitation is shown in figure 1 and is extremely close to that of a constant section Euler-Bernoulli beam. Set E is the elastic modulus, I is the moment of inertia, m is the mass of per unit length, A is the cross-section area and l is the length of the beam. The total displacement of one point on the cantilever beam under the base excitation is defined as:

$$\lambda(x,t)=y(t)+\varphi(x,t) \tag{1}$$

where $\varphi(x,t)$ is the deflection curve, $y(t)$ is the displacement of base. If the base is in sinusoidal movement, it can be defined as $y(t)=Y\sin(\omega t)$, where ω is the radian frequency and Y is the displacement amplitude of the base movement.

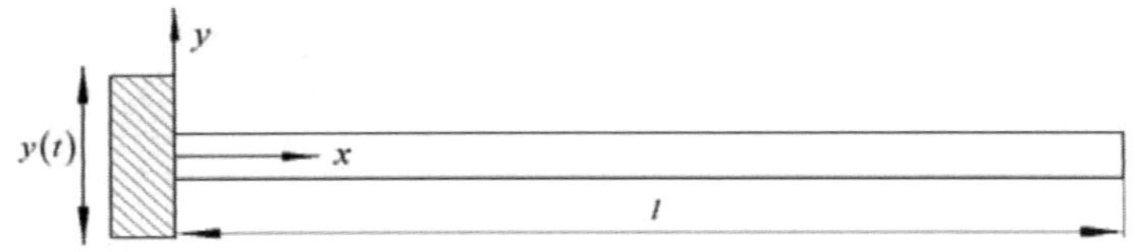

Figure 1. Cantilever beam under base excitation

A small segment is taken from the beam and the force analysis is done which are shown in figure 2. In the figure, V is the shear force on the vertical section, M is the bending moment and c is velocity damping coefficient.

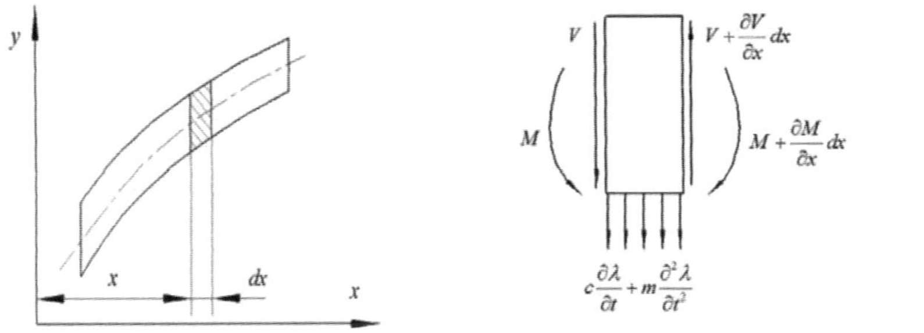

(a) A small segment beam (b) Force analysis on the cross-section

Figure 2. Force analysis

In this study, two types of distributed viscous damping force are taken into consideration [10]: one is external damping force which is proportional to velocity and represented as $D(t)=-c\frac{\partial\lambda}{\partial t}$, the other is internal damping force which is proportional to deformation velocity and expressed as $\sigma_r=-c_r\frac{\partial\varepsilon}{\partial t}$, where c_r is the damping coefficient of deformation velocity and ε is strain. The internal damping force can also produce additional bending moment and shown in figure 3.

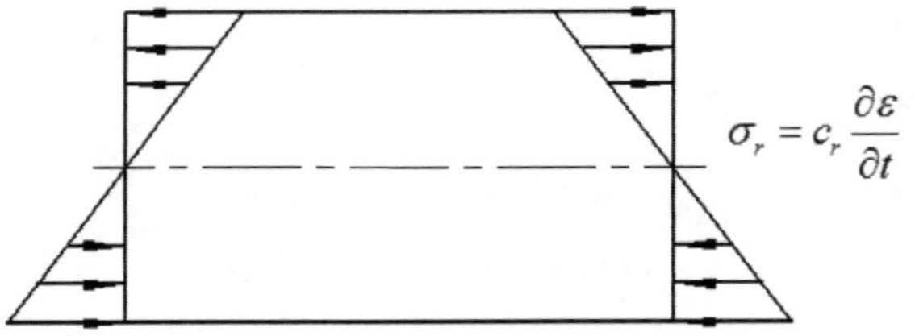

Figure 3. Bending moment produced from the internal damping force

It is assumed that the strain on the cross-section is linear distribution, and then the additional bending moment produced by the internal damping force can be expressed as:

$$M_r=\int_z \sigma_r z\mathrm{d}A=\int_z c_r\frac{\partial\varepsilon}{\partial t}z\mathrm{d}A \tag{2}$$

Where z is the distance between one point and the neutral surface.

From the materials mechanics, the equation between strain and beam deflection can be obtained:

$$\varepsilon=-z\frac{\partial^2\varphi(x,t)}{\partial x^2} \tag{3}$$

Substituting equation (3) into equation (2), the additional bending moment becomes:

$$M_r=\int_z c_r\frac{\partial}{\partial t}\left[-z\frac{\partial^2\varphi(x,t)}{\partial x^2}\right]z\mathrm{d}A=-c_r\frac{\partial^2\varphi(x,t)}{\partial x^2\partial t}\int_z z^2\mathrm{d}A=-c_rI\frac{\partial^2\varphi(x,t)}{\partial x^2\partial t} \tag{4}$$

Then the total bending moment of the small segment is:

$$M=-EI\frac{\partial^2\varphi(x,t)}{\partial x^2}-c_rI\frac{\partial^2\varphi(x,t)}{\partial x^2\partial t} \tag{5}$$

From the moment balance of the whole small segment, can yield:

$$M+\frac{\partial M}{\partial x}\mathrm{d}x-M-V\mathrm{d}x+\frac{1}{2}\left[-c\frac{\partial\lambda(x,t)}{\partial t}-m\frac{\partial^2\lambda(x,t)}{\partial t^2}\right](\mathrm{d}x)^2=0 \tag{6}$$

Neglecting the high order item and referring to the equation (5), the equation (6) can be simplified as:

$$V=\frac{\partial M}{\partial x}=\frac{\partial}{\partial x}\left[-EI\frac{\partial^2\varphi(x,t)}{\partial x^2}-c_rI\frac{\partial^2\varphi(x,t)}{\partial x^2\partial t}\right] \tag{7}$$

Further, from the force balance of the whole small segment, can yield:

$$V+\frac{\partial V}{\partial x}\mathrm{d}x-V-\left[m\frac{\partial^2\lambda(x,t)}{\partial t^2}+c\frac{\partial\lambda(x,t)}{\partial t}\right]\mathrm{d}x=0 \tag{8}$$

Substituting the expression of $\lambda(x,t)$ shown in equation (1), the equation (8) becomes:

$$\frac{\partial V}{\partial x}\mathrm{d}x-\left[m\frac{\partial^2 y(t)}{\partial t^2}+m\frac{\partial^2\varphi(x,t)}{\partial t^2}+c\frac{\partial y(t)}{\partial t}+c\frac{\partial\varphi(x,t)}{\partial t}\right]\mathrm{d}x=0 \tag{9}$$

Further simplification and organization can yield:

$$\frac{\partial V}{\partial x}-m\frac{\partial^2\varphi(x,t)}{\partial t^2}-c\frac{\partial\varphi(x,t)}{\partial t}=m\frac{\partial^2 y(t)}{\partial t^2}+c\frac{\partial y(t)}{\partial t} \tag{10}$$

At last, substituting the expression of V shown in equation (7) into the equation (10) can yield the movement equation of the small segment under the base excitation and shown as:

$$EI\frac{\partial^4\varphi(x,t)}{\partial x^4}+c_r I\frac{\partial^5\varphi(x,t)}{\partial x^4\partial t}+m\frac{\partial^2\varphi(x,t)}{\partial t^2}+c\frac{\partial\varphi(x,t)}{\partial t}=-\left[m\frac{\partial^2 y(t)}{\partial t^2}+c\frac{\partial y(t)}{\partial t}\right] \tag{11}$$

3. Solution of response distributions of cantilever beam under base excitation

It should be noted from the equation (1), if the deflection curve $\varphi(x,t)$ is known, the vibration response of any point on the cantilever beam can be determined.

Here, the modal superposition method was adopted to solve the expression of the deflection curve $\varphi(x,t)$. The every order natural frequency of the beam is set as ω_i, $i=1,2,3,\cdots$, and the relative regular modal shape function is $Y_i(x)$, which satisfies orthogonally conditions as follows:

$$\int_0^l mY_i^2(x)\mathrm{d}x=1 \tag{12}$$

Then the deflection curve $\varphi(x,t)$ can be expressed by the regular response $\eta_i(t)$ and the regular modal shape $Y_i(x)$ and shown as:

$$\varphi(x,t)=\sum_{i=1}^{\infty}Y_i(x)\eta_i(t) \tag{13}$$

For a cantilever beam, the expression of $Y_i(x)$ is:

$$Y_i(x)=D_i\left[\cosh k_i x-\cos k_i x-\frac{\sinh k_i l-\sin k_i l}{\cosh k_i l+\cos k_i l}(\sinh k_i x-\sin k_i x)\right] \tag{14}$$

Where D_i is a parameter related to the excitation level, k_i is a parameter related to nature frequency. Corresponding to the first five order, the values of k_i are $1.875/L$, $4.694/L$, $7.855/L$, $10.996/L$, $14.137/L$ respectively.

Substituting equation (13) into equation (11) can yield:

$$\sum_{i=1}^{\infty} EI\frac{\mathrm{d}^4}{\mathrm{d}x^4}Y_i(x)\eta_i(t)+\sum_{i=1}^{\infty} c_r I\frac{\mathrm{d}^4}{\mathrm{d}x^4}Y_i(x)\dot{\eta}_i(t)+\sum_{i=1}^{\infty} mY_i(x)\ddot{\eta}_i(t)+\sum_{i=1}^{\infty} cY_i(x)\dot{\eta}_i(t) = -\left[m\frac{\partial^2 y(t)}{\partial t^2}+c\frac{\partial y(t)}{\partial t}\right] \tag{15}$$

Utilizing the orthogonally conditions, equation (15) is integrated along the longitudinal direction of the beam and can become:

$$\eta_i(t)\int_0^l EI\frac{\mathrm{d}^4 Y_i(x)}{\mathrm{d}x^4}Y_i(x)\mathrm{d}x+\sum_{i=1}^{\infty}\dot{\eta}_i(t)\int_0^l\left[c_r I\frac{\mathrm{d}^4 Y_i(x)}{\mathrm{d}x^4}+cY_i(x)\right]Y_i(x)\mathrm{d}x+\ddot{\eta}_i(t)\int_0^l mY_i^2(x)\mathrm{d}x=F_i(t) \tag{16}$$

Where $F_i(t)=\int_0^l-\left[m\frac{\partial^2 y(t)}{\partial t^2}+c\frac{\partial y(t)}{\partial t}\right]Y_i(x)\mathrm{d}x$

Assume that the two damping coefficient c and c_r mentioned above are proportional to m and E, and express as $c=\alpha m, c_r=\beta E$. Then equation (16) can be simplified as:

$$\ddot{\eta}_i(t)+\left(\alpha+\beta\omega_i^2\right)\dot{\eta}_i(t)+\omega_i^2\eta_i(t)=F_i(t) \tag{17}$$

Accordingly, $F_i(t)=\int_0^l-\left[m\frac{\partial^2 y(t)}{\partial t^2}+\alpha m\frac{\partial y(t)}{\partial t}\right]Y_i(x)\mathrm{d}x$.

Furthermore, the *i*-order modal damping ratio ζ_i can be expressed as:

$$\zeta_i=\frac{\alpha+\beta\omega_i^2}{2\omega_i}=\frac{\alpha}{2\omega_i}+\frac{\beta\omega_i}{2} \tag{18}$$

Thus the coefficient α, β can be determined from the modal damping ratio and nature frequency, the solution formulas are:

$$\alpha=\frac{2\omega_1\omega_2\left(\zeta_1\omega_2-\zeta_2\omega_1\right)}{\omega_2^2-\omega_1^2} \tag{19}$$

$$\beta=\frac{2\left(\zeta_2\omega_2-\zeta_1\omega_1\right)}{\omega_2^2-\omega_1^2} \tag{20}$$

Where $\zeta_1, \zeta_2, \omega_1, \omega_2$ can be obtained from the experiment.

Combining equation (17) and equation (18) yields the final equation:

$$\ddot{\eta}_i(t)+2\zeta_i\omega_i\dot{\eta}_i(t)+\omega_i^2\eta_i(t)=F_i(t) \tag{21}$$

The solution of equation (21) can be easily obtained by reference to that of a single-degree-of-freedom system and shown as:

$$\eta_i(t) = \mathrm{e}^{-\zeta_i\omega_i t}\left[\frac{\dot{\eta}_{i0}+\zeta_i\omega_i\eta_{i0}}{\omega_i'}\sin\omega_i' t + \eta_{i0}\cos\omega_i' t + \frac{1}{\omega_i'}\int_0^t F_i(\tau)e^{\zeta_i\omega_i\tau}\sin\omega_i'(t-\tau)\mathrm{d}\tau\right] \tag{22}$$

Where

$$\omega_i' = \omega_i\sqrt{1-\zeta_i^2} \tag{23}$$

$$\eta_{i0} = \int_0^l m\varphi(x,0)Y_i\mathrm{d}x \tag{24}$$

$$\dot{\eta}_{i0} = \int_0^l m\dot{\varphi}(x,0)Y_i\mathrm{d}x \tag{25}$$

Finally, the total displacement of the cantilever beam under sinusoidal base excitation is expressed as:

$$\lambda(x,t) = y(t) + \varphi(x,t) = y(t) + \sum_{i=1}^{\infty} Y_i(x)\eta_i(t) \tag{26}$$

From the equation (26), the response distributions of cantilever beam under base excitation can be determined.

4. Study case

4.1. Problem description

In this study, a Ti-6Al-4V beam which is in cantilever state is chosen as study object. The shaking table is taken as base excitation device and the acceleration sensor of light mass is taken as the vibration test device. Then the experiment system is constructed and shown in figure 4 and the relative experimental equipments are listed in table 1 detailedly.

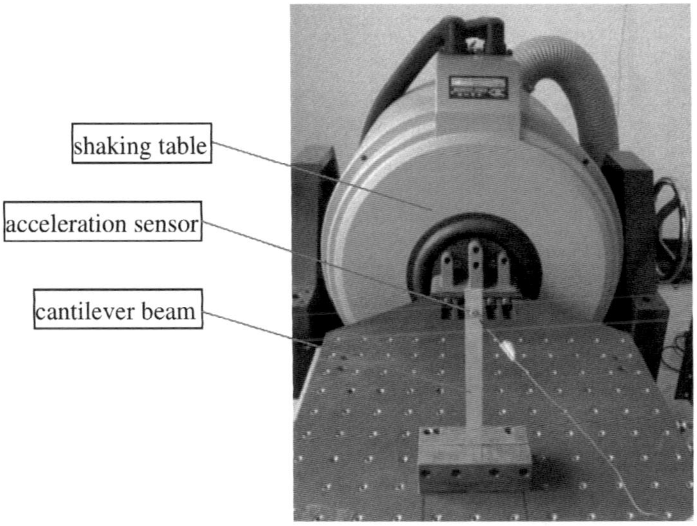

Figure 4. Experiment system

Table 1. The experimental equipments in the base excitation test.

No.	Name
1	LMS SCADAS mobile front-end
2	PCB 8206-001 54627 modal hammer
3	KINGDESIGN EM-1000F shaker
4	B&K4517 acceleration sensor of light mass
5	High-performance notebook computers

The material parameters of the beam have already been tested in the related research. The Young's modulus is 110.32Gpa, Poisson's ratio is 0.31, and the density is $4420\text{kg}/\text{m}^3$. The finished dimensions of the beams are $250\text{mm}\times 20\text{mm}\times 1.5\text{mm}$ and a 25mm section to be mounted in the clamping device.

4.2. Solution of the response of Cantilever Beam by analytical method

During the analytical analysis, the base excitation is set as displacement harmonic excitation with frequencies 239 Hz and 510 Hz respectively and the level of base excitation need be set as consistency with the experiment. Because the level of base excitation is depicted by acceleration in the real experiment, the acceleration should be converted into the displacement by the formula $Y = a/\omega^2$, where a is the acceleration amplitude. In this study, the acceleration excitation amplitudes are set as $0.3g, 0.8g, 1.6g$ respectively. Corresponding to these acceleration amplitudes, the displacement amplitudes which are used for the analytical analysis are listed in table 2.

Table 2. Displacement excitation amplitude $(\times 10^{-7}\text{m})$.

Exciting frequency	Exciting level of shaking table		
	0.3g	0.8g	1.6g
239	13.037	34.767	69.533
510	2.863	7.635	15.270

The damping parameter can be calculated by the modal damping ratios acquired in the subsequent experiment, and the obtained α is 0.518 and β is 3.168×10^{-8}.

The natural frequencies of the cantilever beam are calculated firstly and are listed in table 3.Then equation (26) is applied to compute the displacement response under base excitation with different exciting levels and frequencies. Here, two location on the beam are chosen as response testing points which exist the distances of 0.1m and 0.15m from the clamp area. In order to compare with the experiment, it is required that the displacement responses obtained by analytical solution be transformed into the acceleration responses. Figure 5 is an example about the response of $x = 0.15$ m location. In this calculation, the excitation amplitude is $1.6g$ and excitation frequency is 510Hz. From the figure 5, the response amplitude can be obtained and the value is 13.05m/s^2, Similar with the example, the responses of the beam under different excitation frequency and levels can be computed for the two locations and the relative calculation results are listed in from table 4 to table 7.

Table 3. The natural frequencies by analytical calculation and measurement.

No.	1	2	3	4	5
Calculated value A	23.9	149.8	419.6	822.3	1359.2
Measured value B	22.8	145.4	403.5	789.6	1297.4
Difference $\lvert A-B\rvert / B$	4.82	3.03	3.99	4.14	4.76

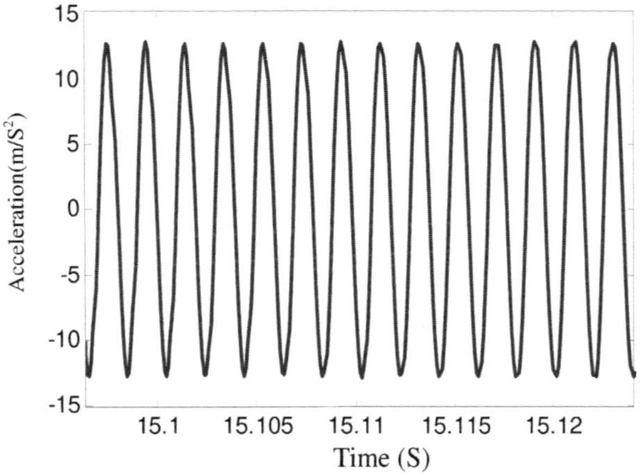

Figure 5. Steady- state response of the cantilever beam

Table 4. Response of the beam when excitation frequency is 239Hz and x=0.1m/ m/s^2.

Exciting level	0.3g	0.8g	1.6g
Calculated value A	1.20	3.19	6.21
Measured value B	1.01	2.68	5.37
Difference $\lvert A-B\rvert / B(\%)$	18.81	19.03	15.64

Table 5. Response of the beam when excitation frequency is 510Hz and x=0.1m/ m/s^2.

Exciting level	0.3g	0.8g	1.6g
Calculated value A	2.43	6.25	12.44
Measured value B	2.78	7.43	14.80
Difference $\lvert A-B\rvert / B(\%)$	12.59	15.88	15.95

Table 6. Response of the beam when excitation frequency is 239Hz and x=0.15m/ m/s^2.

Exciting level	0.3g	0.8g	1.6g
Calculated value A	1.83	5.16	10.21
Measured value B	1.79	4.76	9.55
Difference $\lvert A-B\rvert / B(\%)$	2.24	8.40	6.91

Table 7. Response of the beam when excitation frequency is 510Hz and x=0.15m/ m/s^2.

Exciting level	0.3g	0.8g	1.6g
Calculated value A	2.39	6.57	13.05
Measured value B	2.41	6.46	12.90
Difference $\|A-B\|/B(\%)$	0.83	1.70	1.16

4.3. Response test of the Cantilever Beam.

Corresponding to the analytical analysis, two measuring points are arranged in the cantilever beam. Firstly, the natural frequencies and modal damping ratios are achieved by the hammer exciting method. The natural frequencies obtained by experiment are also listed in table 3 and the modal damping ratios are listed in table 8. Furthermore, the base excitation is done at the frequencies and levels mentioned in analytical analysis to get the responses of two designated points.

Table 8. The modal damping ratios of the beam obtained by experiment.

1	2	3	4	5
0.393	0.035	0.019	0.013	0.022

Similar with analytical analysis, the response of when the excitation frequency is 510Hz and the location is $x = 0.15$ m also is taken as an example of the experiment and shown in figure 6.

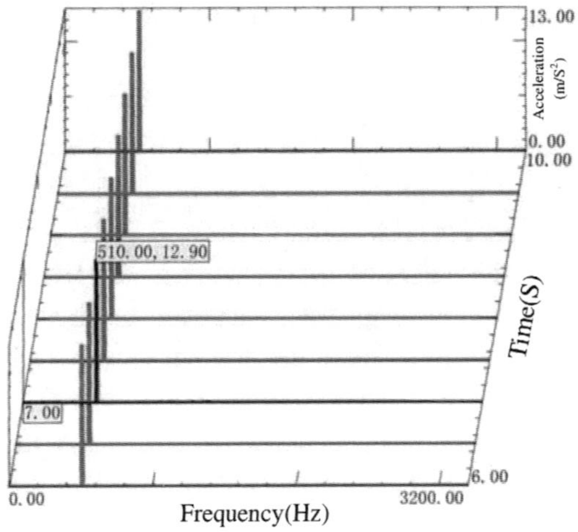

Figure 6. 3D waterfall chart of the steady-state response of the cantilever beam

From the figure 6, the response amplitude can be obtained and the value is 12.90m/s^2 which is almost consistent with the analytic calculation results. The other experiment results are also listed in from table 4 to table 7.

4.4. The Comparison between Experiment and Analysis

From the comparison in the table 4 to table 7, it can be known that there is only a little difference between the experiment and analysis. In particular, the response at $x = 0.15\text{m}$, the maximum difference between the experiment and analysis is smaller than 7% for the different exciting frequencies, which can be thought consistent of both. Then, the correctness of the proposed method about the response analysis of cantilever beam under base excitation is verified. At the location of $x = 0.1$ m, the difference between the analysis and experiment is much bigger than the results of $x = 0.15$m. The bigger difference may be come from the inaccurate position of vibration pickup point. To verify this conclusion, the position of vibration pickup point is modified as $x = 0.096$m and analytical calculation is done again. The relative results are listed in table.9.

Table 9. Response of the beam when excitation frequency is 239Hz.

Exciting level	0.3g	0.8g	1.6g
Calculated value *A*(x=0.096m)	1.00	2.60	5.31
Measured value *B*	1.01	2.68	5.37
Difference $\lvert A-B \rvert / B\,(\%)$	0.99	2.99	1.12

It has been shown in table 9, changing the value of x in the vicinity of $x = 0.1$, the difference between the calculation results and experiment become smaller. This shows that the difference between experiment and analytical analysis at the location of $x = 0.1$ is indeed from the inaccuracy of distance.

5. Conclusion

In this paper the response calculation of cantilever beam under the base excitation was studied. On the basis of considering the characteristic of base excitation and the system damping of cantilever beam, the movement equation of the cantilever beam was derived by moment and force equilibrium. Furthermore, the modal superposition method was adopted to solve the vibration response distributions of the cantilever beam at base excitation. At last, a Ti-6Al-4V beam which is in cantilever state was chosen as study object and the relevant experiment system was set up. The response distributions of cantilever beam under base excitation were computed and compared with the experiment results. The two results are consistent, and then the correctness of the proposed method is verified.

The proposed method can promote the understanding of dynamics mechanism of structure under base excitation and provide a reference for the dynamic analysis of similar structure under base excitation.

6. References

[1] Araki Y, Kawabata S, Asia T and Masui T 2011 *J. Engineering Structures.* **33** 3610
[2] Pellicano F 2011 *J. Solids and Structures.* **48** 408
[3] Han Q, Wang J and Li Q 2011 *J. Sound and Vibration.* **330** 3473
[4] Saranik M, Lenoir D and Jézéquel L 2012 *J. Computers and Structures.* **112-113** 327
[5] Liu J T, Du P A and Huang M 2009 *J. Journal of University of Electronical Science and Technology of China.* **38** 1052
[6] Allahdadian S, Boroomand B and Barekatein A R 2011 *J. Finite Elements in Analysis and Design.* **61** 60

[7] Feng A H, Zhu X D and Lan X J 2011 *J. International Journal of Non-Linear Mechanics.* **46** 1330
[8] Kocatur kT 2004 *J. Sound and Vibration*, **281** 1145
[9] Kumar V, Miller J K and Rhoads J F 2011 *J. Sound and Vibration.* **330** 5401
[10] Clough R W and Penzien 2011 *Dynamics of Structures*(Berkeley: Computers & Structures)

Journal of Physics: Conference Series **448** (2013) 012011 doi:10.1088/1742-6596/448/1/012011

Dynamic analysis of hoisting viscous damping string with time-varying length

P Zhang[1], J H Bao[2] and C M Zhu[3]

[1]Doctor, School of Mechanical Engineering, Shanghai Jiaotong University, Shanghai, People's Republic of China
[2]Doctoral student, School of Mechanical Engineering, Shanghai Jiaotong University, Shanghai, People's Republic of China.
[3]Professor, School of Mechanical Engineering, Shanghai Jiaotong University, Shanghai, People's Republic of China.

E-mail: zhp_roc@sjtu.edu.cn

Abstract. The nonlinear dynamic analysis of a hoisting viscous damping string with time-varying length is investigated. The hoisting string is modeled as a taut translating string with a rigid body attached at its low end. A systematic procedure for deriving the system model of hoisting viscoelastic string with time-varying is presented. The governing equations are developed employing the extended Hamilton's principle considering coupling of axial movement and flexural deformation of string. The Galerkin's method and the 4th Runge-Kutta method are employed to numerically analyze the resulting equations. The motions of elevator hoisting system are presented to illustrate the proposed mathematical models. The results of simulation show that the material viscous damping helps stabilize the transverse vibration. The modeling methods can represent the transverse vibration of hoisting viscous damping string with time-varying length.

1. Introduction

While rope is employed in hoisting industry such as mine hoists, elevators, cranes etc, it is subject to vibration due to its high flexibility and relatively low internal damping characteristics [1-3]. Most often these systems are modeled as either an axially moving tensioned beam or string with time-varying length and a load at its lower end [4, 5]. It was shown that the vibration energy of the rope changes in general during elongation and shortening. When the rope length is being shortened, vibration energy increases exponentially with time, causing dynamic instability [6]. So the vibration of the rope with time-varying length has been studied by some researchers.

Kotera and Kawai [7] analyzed free vibrations of a string with time-varying length and a weight at one end. New variables of position and time are introduced which allow the equation of motion to be solved by Laplace transformation. Almost exact solutions of free vibrations induced by a distributed initial displacement are also obtained. Kaczmarczyka and Ostachowiczb [8] studied coupled vibration of deep mine hoisting cable and built a distributed-parameter model. They found that response of the catenary-vertical rope system may feature a number of resonance phenomena. Kimure and Iijima et al. [9-12] published a series of studies on vibrations of elevator rope. Finite difference analyses of the rope vibration were performed by considering the time-varying length of the rope, based on the assumption that the movement velocity is constant. The calculated results of the finite difference analyses are in fairly good agreement with the calculated results of the exact solution. The above studies assume that the velocity is constant during the process of modeling. Some researches [13-15] assumed that the transport speed is characterized as a small simple harmonic variation about the constant mean speed. The assumption has its physical meaning. For example, if the axially moving rope with time-varying length models a string on a rotating pulley, the rotation vibration of the pulley will result in a small fluctuation in the axial speed of the string. However, the study of vibrations of varying length rope with arbitrary varying axial velocity is a relatively less studied problem in literature. Zhu and Chen [16] investigated a comprehensive, theoretical and experimental study of the uncontrolled and controlled lateral responses of a moving cable in a high-rise elevator. A novel experimental method was developed to validate the uncontrolled and controlled response and shown good agreement with the theoretical predictions. Zhang and Zhu et al. [17] derived the governing equation and energy equation of longitudinal vibration of flexible hoisting system with arbitrarily varying length and velocity.

Extensive research efforts on the flexible hoisting rope with time-varying length have been done in the last few decades as aforementioned, however, most studies assumed the rope is linearly elastic, and damping was ignored since the primary focus was on the intrinsic stability. In the paper, the dynamic characteristics of the hoisting viscous damping rope with time-varying length are the subject of this investigation. The governing equations are developed employing the extended Hamilton's principle. The derived governing equations are shown to be partial differential equations (PDF) with variable coefficients. On choosing proper mode functions that satisfy the boundary conditions, the solution of the governing equations was obtained using the Galerkin's method. The motions of elevator hoisting rope were illustrated to evaluate the proposed mathematical models. According to the numerical simulations, the effects of the viscous damping of material on the dynamic characteristics are analyzed for the hoisting rope with time-varying length. Based on the proposed dynamic analysis, further vibration control will be adopted for such the hoisting systems in the near future.

2. Model of hoisting system

Hoisting system can be simplified as an axially moving string with time-varying length and a rigid body m at its lower end, as shown in Figure 1. The rail and the suspension of the rail are assumed to be rigid. The string has Young's modulus E, diameter d and the density per unit length ρ. The origin of coordinate is set at the top end of string and the instantaneous length of string is $l(t)$ at time t. The instantaneous translational velocity, acceleration and jerk of the string are $v(t)$, $a(t)$ and $j(t)$ respectively. At any instant t, the transverse displacements of string is described by $y(x,t)$, at a spatial position $x(t)$, where $0 \le x(t) \le l(t)$. The following assumptions constrain the analysis: 1. Young's modulus E, diameter d and density ρ of the string are always constants; 2. Only transverse vibration is considered here. The elastic distortion of string arousing from the transverse vibration is much less than the length of the string.

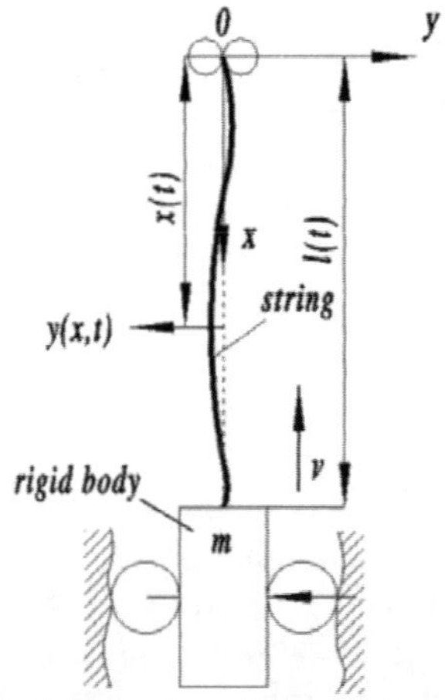

Figure 1. Schematic of hoisting viscoelastic string with time-varying length

The kinetic energy of flexible hoisting system is computed by

$$E_k(t) = 0.5(m + \rho l)v^2 + 0.5my_t^2(l(t),t) + 0.5\rho\int_0^{l(t)} (y_t(x,t) + vy_x(x,t))^2 dx \tag{1}$$

The elastic strain energy of the string is

$$E_e(t) = \int_0^{l(t)} \left(P\varepsilon + \frac{1}{2}EA\varepsilon^2 \right)dx + \frac{1}{2}\int_0^{l(t)} EIy_{xx}^2 dx \tag{2}$$

The first and the second terms on the right of Equation (2) represent axial strain energy of the string, the third term represents bending strain energy of the string. $P(x,t)$ is the quasi-static tension in the string and is given by

$$P = [m + \rho(l(t) - x)]g \tag{3}$$

And ε represents the strain measure at spatial position x of the string and can be expressed as

$$\varepsilon = (\mathrm{d}s - \mathrm{d}x)/\mathrm{d}x \tag{4}$$

As shown in Figure 2, ds can be expressed as

$$\mathrm{d}s \approx \sqrt{1+(\mathrm{d}y/\mathrm{d}x)^2}\,\mathrm{d}x \approx \left[1+\frac{1}{2}\left(\frac{\partial y}{\partial x}\right)^2 - \frac{1}{8}\left(\frac{\partial y}{\partial x}\right)^4 + \cdots\right]\mathrm{d}x \approx \left[1+\frac{1}{2}\left(\frac{\partial y}{\partial x}\right)^2\right]\mathrm{d}x \tag{5}$$

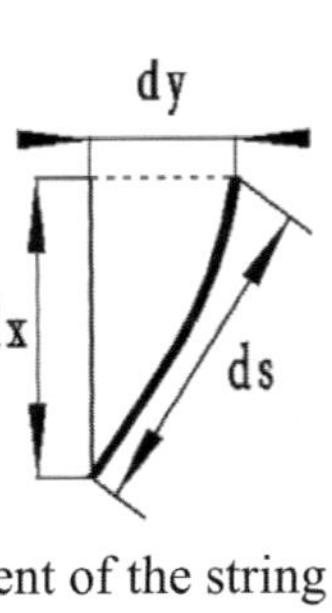

Figure 2. A small element of the string in a deformed position

Substituting Equation (5) into Equation (4) yields

$$\varepsilon = \frac{1}{2}y_x^2 \tag{6}$$

The virtual work done by the transverse damping forces of the string is given by

$$\delta W = \int_0^{l(t)} F_c \delta y dx \tag{7}$$

where F_c is the viscous damping distributed force[2] in the string, and

$$F_c = -c_1 P y_{xxt} + c_2 \rho y_t \tag{8}$$

where c_1 and c_2 are constant coefficients of transverse damping. According to the characteristics of top restriction of the string, the boundary conditions at $x(t)=0$ are

$$y(0,t) = 0\,,\ \ y_t(0,t) = 0 \tag{9}$$

On substitution of Equs.(1), (2) and (7) in the extended Hamilton's Principle,

$$\delta\int_{t_1}^{t_2}(E_k(t) - E_e(t))dt + \int_{t_1}^{t_2}\delta W dt = 0 \tag{10}$$

and apply the variational operation and the procedure for integration by parts to obtain,

$$\int_{t_1}^{t_2}\left[m\frac{\partial}{\partial t}y_t(l,t)+\left(P+\frac{1}{2}EAy_x^2\right)y_x - EIy_{xxx}\right]\delta y\Big|_l dt + \int_{t_1}^{t_2}\left(EIy_{xx}\right)\delta y_x\Big|_l \mathrm{d}t +$$
$$\int_{t_1}^{t_2}\int_0^l\left[\rho\frac{\partial}{\partial t}(y_t+vy_x)+\rho v\frac{\partial}{\partial x}(y_t+vy_x)-\frac{\partial}{\partial x}\left(\left(P+\frac{1}{2}EAy_x^2\right)y_x\right)+F_c+EIy_{xxxx}\right]\delta y\mathrm{d}x\mathrm{d}t = 0 \tag{11}$$

Setting the coefficients of δ_y in Equation (11) to zero yields the governing equation for the string,

$$\rho(y_{tt}+2vy_{xt}+ay_x+v^2y_{xx}) - P_x y_x - P y_{xx} - \frac{3}{2}EAy_x^2 y_{xx} - c_1 P y_{xxt} + c_2\rho y_t + EIy_{xxxx} = 0, 0<x<l(t) \tag{12}$$

The first four terms in Equation (12) correspond to the local, Coriolis, tangential, and centripetal acceleration, respectively. Equation (12) is a partial differential equation which describes the dynamics of the flexible hoisting string. The equation defined over time-dependent spatial domain rendering the problem non-stationary. Hence, the exact solution to this problem is not available, and recourse must be made to an approximate analysis. In what follows, numerical techniques are employed to obtain approximate solution for the governing equation.

3. Discretization of the governing equation

Equation (12) is a partial differential equation with infinite dimensions and many parameters are time-variant. It is impossible to obtain an exact analytical solution from Equation (12). In this section, Galerkin's method is applied to truncate the infinite-dimensional partial differential equation into a linear finite-dimensional ordinary differential equation with time-variant coefficients. Then, solve them with numerical methods. In order to map Equation (12) onto the fixed domain, a new independent variable $\zeta=x/[l(t)]$ is introduced and the time-variant domain $[0, l(t)]$ for x is converted to a fixed domain [0, 1] for ζ. According to the characteristic of taut translating string, the solution of $y(x, t)$ is assumed in the forms[5,13]

$$y(x,t) = \sum_{i=1}^{n}\varphi_i(\zeta)q_i(t) = \sum_{i=1}^{n}\varphi_i\left(\frac{x}{l}\right)q_i(t) \tag{13}$$

where $q_i(t)$ (i=1,2,3,…,n) is the generalized coordinate respect to $y(x,t)$, n is the number of included mode. $\varphi_i(\zeta)$ is trial function[5,13],

$$\varphi_i(\zeta)=\sqrt{2}\sin i\pi\zeta \tag{14}$$

Consequently, expansion Equation (13) results in the expressions for partial derivatives of transverse displacement functions,

$$y_x(x,t)=\frac{1}{l(t)}\sum_{i=1}^{n}\phi_i'(\zeta)q_i(t),\quad y_{xx}(x,t)=\frac{1}{l(t)^2}\sum_{i=1}^{n}\phi_i''(\zeta)q_i(t)$$

$$y_{xxxx}(x,t)=\frac{1}{l(t)^4}\sum_{i=1}^{n}\phi_i^{(4)}(\zeta)q_i(t),$$

$$y_{xt}(x,t)=\sum_{i=1}^{n}\frac{1}{l(t)}\phi_i'(\zeta)\dot{q}_i(t)-\sum_{i=1}^{n}\left(\frac{\zeta v(t)}{l(t)^2}\phi_i''(\zeta)+\frac{v(t)}{l(t)^2}\phi_i'(\zeta)\right)q_i(t),$$

$$y_{tt}(x,t)=$$

$$\sum_{i=1}^{n}\phi_i(\zeta)\ddot{q}_i(t)-\frac{2\xi v(t)}{l(t)}\sum_{i=1}^{n}\phi_i'(\zeta)\dot{q}_i(t)+\sum_{i=1}^{n}\left(\frac{2\zeta v(t)^2}{l(t)^2}\phi_i'(\zeta)-\frac{\zeta a(t)}{l(t)}\phi_i'(\zeta)+\frac{\zeta^2 v(t)^2}{l(t)^2}\phi_i''(\zeta)\right)q_i(t) \tag{15}$$

where the overdot denotes time differentiation. Substituting Equation (13)-(15) into Equation (12), multiplying the governing equation by $\varphi_k(\zeta)$ (k=1,2,3,…,n), integrating it from ζ=0 to 1, and using the boundary conditions and the orthonormality relation for $\varphi_i(\zeta)$ yield the discretized equation of transverse vibration for the flexible hoisting string with time-variant coefficients

$$\mathbf{M}\ddot{Q}+\mathbf{C}\dot{Q}+\mathbf{K}Q+S(Q)=0 \tag{16}$$

where $Q=[q_1(t),q_2(t),\cdots,q_n(t)]^T$ is vector of generalized coordinate, M, C and K are matrices of mass, damp and stiffness respect to Q, respectively. $S(Q)$ is higher order item of generalized coordinate. The overdot denotes time differentiation. Where entries of the matrices are expressed as follows,

$$M_{ik}=\rho\delta_{ik}$$

$$C_{ik}(t)=\frac{2\rho v}{l}\int_0^1(1-\zeta)\phi_i'(\zeta)\phi_k(\zeta)\mathrm{d}\zeta-\frac{c_1\rho g}{l}\int_0^1(1-\zeta)\phi_i''(\zeta)\phi_k(\zeta)\mathrm{d}\zeta-\frac{c_1 mg}{l^2}\int_0^1\phi_i''(\zeta)\phi_k(\zeta)\mathrm{d}\zeta+c_2\rho\delta_{ik}$$

$$K_{ik}(t)=\frac{\rho a}{l}\int_0^1(1-\zeta)\phi_i'(\zeta)\phi_k(\zeta)\mathrm{d}\zeta-\frac{\rho v^2}{l^2}\int_0^1(1-\zeta)^2\phi_i'(\zeta)\phi_k(\zeta)\mathrm{d}\zeta+\frac{\rho g}{l}\int_0^1(1-\zeta)\phi_i'(\zeta)\phi_k'(\zeta)\mathrm{d}\zeta-$$

$$\frac{mg}{l^2}\int_0^1\phi_i''(\zeta)\phi_k(\zeta)\mathrm{d}\zeta+\frac{EI}{l^4}\int_0^1\phi_i^{(4)}(\zeta)\phi_k(\zeta)\mathrm{d}\zeta+$$

$$\frac{2vc_1\rho g}{l^2}\int_0^1(1-\zeta)\phi_i''(\zeta)\phi_k(\zeta)\mathrm{d}\zeta+\frac{vc_1\rho g}{l^2}\int_0^1\zeta(1-\zeta)\phi_i'''(\zeta)\phi_k(\zeta)\mathrm{d}\zeta+\frac{2vc_1 mg}{l^3}\int_0^1\phi_i''(\zeta)\phi_k(\zeta)\mathrm{d}\zeta+$$

$$\frac{vc_1 mg}{l^3}\int_0^1\zeta\phi_i'''(\zeta)\phi_k(\zeta)\mathrm{d}\zeta-\frac{c_2\rho v}{l}\int_0^1\zeta\phi_i''(\zeta)\phi_k(\zeta)\mathrm{d}\zeta$$

$$S_k(\mathbf{Q})=-\frac{3EA}{2l^4}\int_0^1\left(\sum_{i=1}^{n}\phi_i'(\zeta)q_i(t)\right)^2\sum_{i=1}^{n}\phi_i''(\zeta)q_i(t)\phi_k(\zeta)\mathrm{d}\zeta \tag{17}$$

where the superscript "'" denotes partial differentiation for normalized variable ζ, δ_{ij} is the Kronecker delta defined by $\delta_{ik}=1$ if $i=k$ and $\delta_{ik}=0$ if $i\neq k$ ($i=1,2,3,\ldots,n$, $k=1,2,3,\ldots,n$). If the initial displacement and velocity of the string are given by $y(x,0)$ and $y_t(x,0)$, respectively, where $0<x<l(0)$, the initial conditions for the generalized coordinate can be obtained from Equation (13) and (15),

$$q_i(0)=\int_0^1 y(\zeta l(0),0)\phi_i(\zeta)\mathrm{d}\zeta \tag{18}$$

$$\dot{q}_i(0)=\int_0^1 y_t(\zeta l(0),0)\phi_i(\zeta)\mathrm{d}\zeta+\frac{v(0)}{l(t)}\sum_{i=1}^{n} q_i(0)\int_0^1 \zeta\phi_i'(\zeta)\phi_k(\zeta)\mathrm{d}\zeta \tag{19}$$

Solving the ordinary differential Equation (16) with numerical methods may yield the instantaneous values of Q. Substituting these values into Equation (13) may yield the instantaneous values of transverse vibration of the string $y(x, t)$.

4. Numerical simulation and discussions

Equation (16) is a second-order ordinary differential equations with variable coefficients. They will be used to investigate the dynamic responses of hoisting viscous damping string with time-varying length. The solutions of the equation are obtained by the Runde-Kutta method. Kimura [9-12] published a series of studied on vibrations of a string with time-varying length. However, these studies considered the string with uniform motions only. In the study, we will consider the speed changes with time. In what follows, the motions of elevator hoisting system were illustrated to evaluate the proposed mathematical model. Elevator hoisting system is simplified as an axially translating viscous damping string with a rigid body attached at its lower end. The simulation parameters for the elevator are given in Table 1. The flight time for a travel distance of 135m (45 stories) is 33 second. Figure 3 gives the prescribed displacement, velocity, acceleration and jerk curves of elevator hoisting system. Utilizing the curves as the input of Equation (16) with aid of MatLab may obtain dynamic responses of elevator hoisting system. In this work, all numerical analyses were implemented with aid of MatLab.

Table 1. Simulation parameters

Items	Data values
Density per unit length ρ(kg/m)	0.707
Young's modulus E (N/m^2)	2.01×1011
String diameter d(m)	14×10^{-3}
Hoisting mass m(kg)	300
Minimum length of the string lmin (m)	5
Maximum length of the string lmax (m)	140
Maximum velocity vmax(m/s)	5
Maximum acceleration amax(m/s^2)	1
Maximum jerk jmax(m/s^3)	1
Total travel time t(s)	33
Number of transverse modes n	4

Consider the free vibration caused by a distributed initial displacement and released from rest. The initial displacement and velocity are respectively

$$y(x,0)=y_0\sin\frac{x\pi}{l_0},\quad y_t(x,0)=0 \tag{20}$$

where y_0=0.005m is the initial amplitude. Transverse vibration responses of hoisting viscous damping string at 2m above the car during movement of elevator are illustrated in Figure 4 and Figure 5.

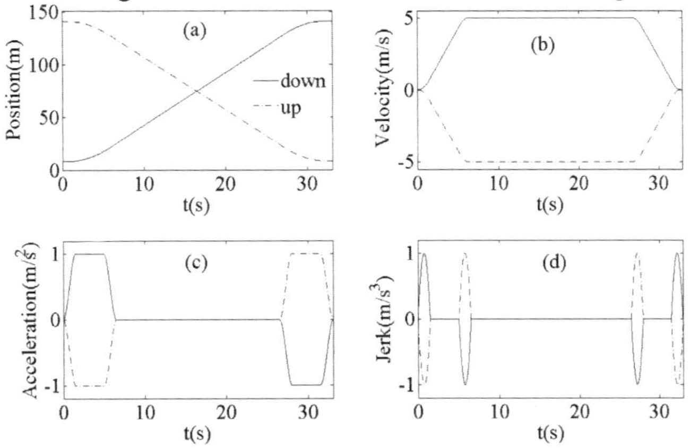

Figure 3. Movement profile of the elevator: (a)$l(t)$; (b)$v(t)$;(c)$a(t)$; (d)$j(t)$.

Consider the free vibration caused by a distributed initial displacement and released from rest. The initial displacement and velocity are respectively

$$y(x,0) = y_0 \sin\frac{x\pi}{l_0}, \quad y_t(x,0) = 0 \tag{20}$$

where y_0=0.005m is the initial amplitude. Transverse vibration responses of hoisting viscous damping string at 2m above the car during movement of elevator are illustrated in Figure 4 and Figure 5.

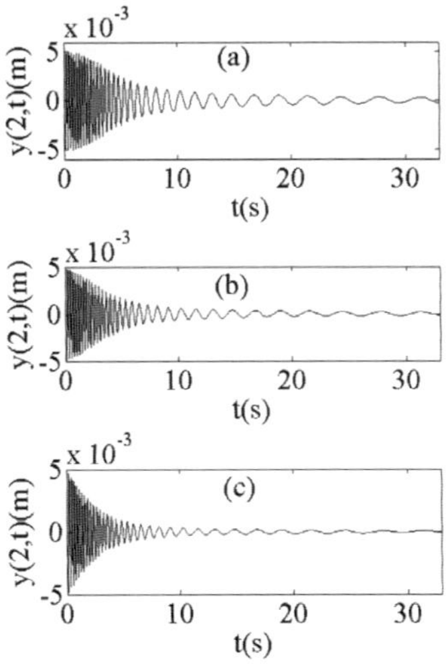

Figure 4. Transverse vibration responses of hoisting viscous damping string at 2m above the car during downward movement of the elevator: (a)c_1=0; (b) c_1=0.0003; (c) c_1=0.0008.

Figure 4 displays reducing vibration amplitudes with increasing length of the rope during downward movement. This is due to the energy of flexible hoisting system transfers from the transverse vibration to the axial motion by bringing some mass into the domain of effective length, i.e., the axially hoisting rope is dissipative during downward movement, thus leading to a stabilized transverse dynamic response. A possible physical interpretation of the result is as follows: during downward movement negative external work is required to maintain the prescribed axial motion which, in turn, brings about a convection of mass in the domain of effective length. At the same time, frequencies of the transverse vibration reduce with increasing length of the rope. This is due to the fact that the mass of the rope increase and the stiffness of the rope decrease, i.e., the rope becomes somewhat "softer".

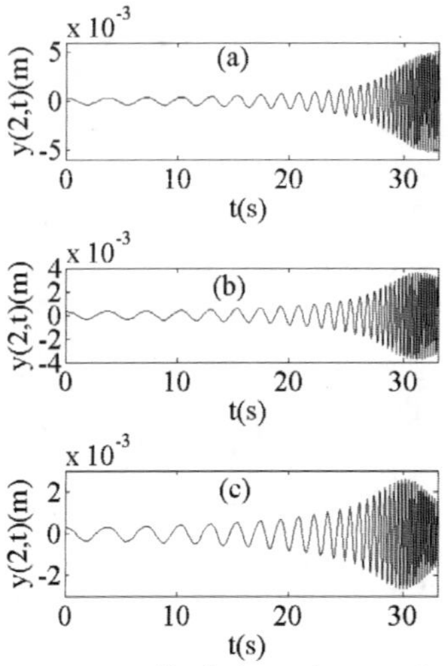

Figure 5. Transverse vibration responses of hoisting viscous damping string at 2m above the car during upward movement of the elevator: (a) c_1=0; (b) c_1=0.0003; (c) c_1=0.0008.

By contrast, in Figure 5, we observe that vibration amplitudes of the rope increase with decreasing length of the rope during upward movement. This is due to the energy of flexible hoisting system transfers from the axial motion to the transverse vibration by leaving some mass out of the domain of effective length, i.e., the axially hoisting rope gains energy during upward movement, thus leading to an unstabilized transverse dynamic response. A possible physical interpretation of the result is as follows: during upward movement positive external work is required to maintain the prescribed axial motion which, in turn, brings about a convection of mass out of the domain of effective length. In the mean time, frequencies of the transverse vibration increase with decreasing length of the rope. This is due to the fact that the mass of the rope decrease and the stiffness of the rope is increase, i.e., the rope becomes somewhat "stiffer".

Further, Figure 4 and Figure 5 show that the viscous damping takes effect on the vibration amplitude. The response amplitude decreases as the viscous damping increases. Higher viscous damping leads to amplitude reduction faster, as shown Figure 4(c) and Figure 5(c). The reduced response amplitude indicates that the material viscous damping makes the dynamic system more stable.

5. Conclusion

The nonlinear vibration characteristics for a flexible hoist string with time-varying length considering coupling of axial movement and flexural deformation is analyzed in this paper. The flexible hoisting system is modeled as an axially moving string with time-varying length and a load m at its lower end. The governing equations are derived by using Leibnitz's rule and the extended Hamilton's principle. The Galerkin's method is used to truncate the infinite-dimensional partial differential equations into a set of nonlinear finite-dimensional ordinary differential equations with time-variant coefficients. Based on the numerical simulations, the following conclusions can be obtained: The natural frequencies of flexible hoisting string with time-varying length are increasing because of the reducing mass and the increasing stiffness of the string, and the energy transforms from the axial movement into the flexible deformation. By contrast, the natural frequencies are decreasing because of the increasing mass and the reducing stiffness of the string, and the energy coverts from the flexible deformation into the axial movement. The material viscous damping always dissipates energy and helps stabilize the transverse vibration during movement of the hoisting system. The proposed the theoretical model in this paper will be helpful for the researchers to comprehend its dynamic behavior and develop the proper method to suppress the vibration in practice.

References

[1] Kaczmarczyk S 1997 *J. Sou. Vib.* **208** 243-69
[2] Kaczmarczyk S and Ostachowicz W 2003 *J. Sou. Vib.* **262** 219-44
[3] Kaczmarczyk S and Andrew P 2005 *Elev. Worl.* 126-29
[4] Zhu W D and Xu G Y 2003 *J. Sou. Vib.* **263** 679-99
[5] Zhu W D and Chen Y 2005 *Multi-body Dyna.* **219** 35-53
[6] Lee S Y and Lee M 2002 *J. Appl. Phys.* **69** 83-87
[7] Kotera T and Kawai R *JSME int. ser.* III, *Vib. Con. Eng. . Ind.* **31** 524-29
[8] Kaczmarczyk S and Ostachowicz W 2003 *J. Sou. Vib.* **262** 245-89
[9] Kimura H and Nakagawa T 2007 *J. Env. Eng.* **2** 76-86
[10] Kimura H, Ito H and Nakagawa T 2007 *J. Env. Eng.* **2** 87-96
[11] Kimura H, Ito H and Nakagawa T 2008 *J. Env. Eng.* **2** 540-49
[12] Kimura H, Iijima T, Matsuo S and Fujita Y 2009 *J. Sys. Des. Dyn.* **3** 420-28
[13] Fung R F, Wang P H and Lee M J 1998 *J. Chin. Ins. Eng.* **21** 109-17
[14] Yao C M, Fung R F and Tseng C D 1999 *J. Sou. Vib.* **219** 323-37
[15] Wang P H, Fung R F and Lee M J 1998 *J. Sou. Vib.* **209** 223-49
[16] Zhu W D and Chen Y 2006 *J. Vib. Aco.* **128** 66-78
[17] Zhang P, Zhu C M and Zhang L J 2008 *J. Shanghai JiaoTong Uni.* **42** 481-88

Effects of the lateral propagating disturbance on wave propagation in periodic beams

Chen Tao and Wang Ligang

College of Science, Harbin Engineering University, Harbin, PR China

E-mail:chen_tao1204@163.com

Abstract. The elastic wave propagation in periodic beams with the propagating disturbance is investigated. The effects of the disturbance on the wave propagation in periodic beams are studied. A propagating wave (disturbance) is incident upon the discontinuity and gives rise to transmitted and reflected waves. All of the transmitted and reflected waves of given flexural wave incident upon the beam at some location are found and superposed using the multiple reflections approach. The relation between the wave-field of incident wave (disturbance) and the wave-field of resulting waves on any segment is expressed. Much attention is devoted to the response in the frequency ranges with gaps in the band structure for the corresponding periodic beams with the disturbance. The numerical results of the frequency response function of finite periodic beams with propagating disturbance are presented. With the increase disturbance amplitude, the attenuation of the band gaps gradually decreases. The effects of the number of cells and material parameters on the band gap are also considered.

1. Introduction

In the last decade much attention is focused on periodic flexural systems that may be idealized as consisting of identical elements connected end to end before [1-5]. The work on phononic band gaps has led to a renewed interest in elastic wave propagation in periodic materials or structures, called phononic crystals, which are made of two or more elastic materials with large contrast between their mechanical properties in recent years [6-8]. The properties of frequency band gaps in the phononic crystals have been studied in both theoretical predictions and experiments [9-14]. Most of the previous studies are focused on the band structure in periodic materials with infinite dimensions. Many potential applications of periodic structures may be expected such as vibration isolation technology and acoustic filters.

As we know, many practical engineering structures are conceived as assemblies of relatively slender elements that can be models as one-dimensional mechanical waveguides. Periodic structures are important both as fundamental structural elements and as simple global models for some slender structures, and the length of such periodic materials or structures in practical engineering structures is finite. Jensen [15] investigated phononic band gaps of finite one-dimensional mass-spring periodic structures subjected to axial loading, and indicated that the frequency response function may be calculated to describe the band gap characteristics of finite periodic structures. Kobayashi et al. [16] studied wave transmission characteristics in one-dimensional multilayered media with a finite length using the transform matrix method. Xiang and Shi [17] determined the band gaps in finite periodic beams using differential quadrature method. Yu et al. [18] investigated the propagation properties of flexural wave in periodic beams using the transform matrix method.

Journal of Physics: Conference Series **448** (2013) 012012 doi:10.1088/1742-6596/448/1/012012

Although many methods may be applied to investigate the band gap of periodic structures, it is very difficult to study effects of lateral disturbances on wave propagation in the periodic beam using existing methods, such as transfer matrix method. These methods may be applied to analyze the band gap characteristic of periodic structure, however, the relation between the wave-field of incident wave (disturbance) and the wave-field of resulting waves on any segment can not be expressed. It is not convenient for their promising applications in vibration isolation and other engineering application. Further investigation on the wave propagation in finite dimensional structures with one or more lateral disturbance is of great importance. Particularly, Ungar [5] first proposed the multiple reflections approach. The approach can set up the relation between the wave-field of incident wave and the wave-field of resulting waves on any segment. However, the approach does not be explained in detail, and the validity of the multi-reflection approach is also not proved numerically.

To our knowledge, no work on effects of lateral disturbances on wave propagation in periodic beams structures has been reported. The present paper aims to analyze effects of lateral disturbances on wave propagation in periodic beams from theoretical as well as numerical points of view. In Section 2, periodic binary beams are adopted, and the results of injecting a propagating wave are concerned. All of the transmitted and reflected waves of flexural wave incident upon the beam at some specified location are found and superposed using the multi-reflection approach. In Section 3, the relation of the wave-field on adjacent identical elements is given, comparing the propagating constant calculated by the relation with calculated by the transfer matrix method, iteration times of wave-field can be determined. In Section 4, the influences of the number of cells and material parameters on the band gap of periodic beams are analyzed. Frequency response of periodic beams can be changed by tuning the disturbance. The conclusions from this study are listed in Section 5. From the result, we can observe some interesting and valuable physical phenomena.Another section of your paper

2. Wave reflection and transmission at discontinuities

As shown in Fig. 1, a periodic binary beam consists of an infinite repetition of alternating segments A with length a_1 and segments B with length a_2. So the periodic beam's lattice constant is $a = a_1 + a_2$. In other word, the whole beam is split into successive unit cells. Each cell contains two segments with length a_1 and a_2. The uniform Euler-Bernoulli beam model with constant cross-section is used.

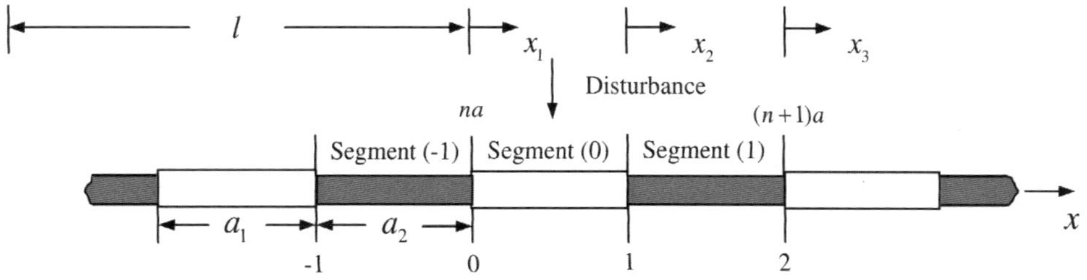

Fig. 1 Designation used for wave multiple-reflection and transmission analysis of a periodic binary beam with a propagating disturbance.

2.1 Multiple reflections between two discontinuities

As shown in Fig. 1, let an external disturbance generate a harmonic wave $w_{in}(x,t)$ on segment (0) and propagating in the positive x direction. The positive-going wave produced will give rise to transmitted and reflected waves when it impinges on discontinuity (1). As the beginning to dealing with more complicated case, one may first study the wave-field set up between two discontinuities due to multiple reflections. The incident wave is described by

$$w_{in}(x) = A_{in1}\exp(-k_1 x\mathrm{i}), \tag{1}$$

where $k_1 = \sqrt[4]{(\rho S)_1 \omega^2 /(EI)_1}$, E denotes Young's modulus, I the area moment of inertia, ρ is the density and S the cross-sectional area, A_{in1} is the wave amplitude.

For simplifying the analysis, the present section deals with the hypothetical case where near-field effects are neglected. Then at the discontinuity (1), a transmitted component $w_{tr1}(x)$ and a reflected component $w_{re1}(x)$ are generated

$$w_{tr1}(x) = A_{tr1}\exp(-k_2 x\mathrm{i}), \tag{2}$$

$$w_{re1}(x) = A_{re1}\exp(-k_1 x\mathrm{i})\exp(2k_1 x_2\mathrm{i}) = A_{re1}\tau\eta\exp(k_1 x\mathrm{i}), \tag{3}$$

where $k_2 = \sqrt[4]{(\rho S)_2 \omega^2 /(EI)_2}$, $x_2 = x - l - a_1$, $\tau = \exp(-2k_1 l\mathrm{i})$, $\eta = \exp(-2k_1 a_1\mathrm{i})$, $A_{re1} = RA_{in1}$, $A_{tr1} = TA_{in1}$, R and T are the transmission and reflection coefficients at the discontinuity (1) of the segment (0), l is the distance between the origin of the x system and the discontinuity (0).

The reflected portion $w_{re1}(x)$ will travel in the negative x direction until it is incident upon discontinuity (0), and there will be transmitted and reflected. The reflected portion

$$w_{re2}(x) = A_{re11}\exp(-k_1 x\mathrm{i})\exp(2k_1 x_2\mathrm{i})\exp(-2k_1 x_1\mathrm{i}) = A_{re11}\eta\exp(-k_1 x\mathrm{i}), \tag{4}$$

where $x_1 = x - l$, $A_{re11} = RA_{re1}$.

The reflected waves $w_{re2}(x)$ will again travel in the positive x direction, and will suffer the same situation as the initial incident waves. So two set of waves will be result in the region between the two continuities. One consists of an infinite number of portions traveling in the positive x direction; the other consists of a same number traveling in the negative x direction. Multiple reflections produce a total wave $w_{na+}^{(0)}(x)$ and a total wave $w_{na-}^{(0)}(x)$ on segment (0) by superposition, where

$$\begin{aligned} w_{na+}^{(0)}(x) &= A_{in1}\exp(-k_1 x\mathrm{i}) + R^2\eta A_{in1}\exp(-k_1 x\mathrm{i}) + R^4\eta^2 A_{in1}\exp(-k_1 x\mathrm{i}) + \cdots \\ &= \frac{1}{1-R^2\eta}A_{in1}\exp(-k_1 x\mathrm{i}) = A_{n1}^{(0)}\exp(-k_1 x\mathrm{i}), \end{aligned} \tag{5}$$

$$\begin{aligned} w_{na-}^{(0)}(x) &= R\tau\eta A_{in1}\exp(k_1 x\mathrm{i}) + R^3\tau\eta^2 A_{in1}\exp(k_1 x\mathrm{i}) + R^5\tau\eta^3 A_{in1}\exp(k_1 x\mathrm{i}) + \cdots \\ &= \frac{R\tau\eta}{1-R^2\eta}A_{in1}\exp(k_1 x\mathrm{i}) = A_{n2}^{(0)}\exp(k_1 x\mathrm{i}), \end{aligned} \tag{6}$$

where the superscript (0) is added to denote the initial approximations. Periodic beam structures is considered, so the modulus of $R^2\eta$ is less than 1.

2.2 Unidirectional propagation

Above Section neglected transmitted waves across the boundaries of segment (0) in Fig. 1. It is difficult to account for all of the transmitted waves of two directions using the method of multiple reflections at the same time, so wave-field will be first set up when wave transmission takes place in only one coordinate direction.

If waves are transmitted in only rightward direction, then wave-field established on segment (0) $w_{na+}^{(0)}(x)$ will be transmitted past discontinuity (1), and the total transmission will suffer multiple reflections between discontinuity (1) and (2). Multiple reflections produce a total wave $w_{nb+}^{(0)}(x)$ and a total wave $w_{nb-}^{(0)}(x)$ on segment (1) by superposition, where

$$w_{nb+}^{(0)}(x)=\frac{T}{1-R^2\eta}A_{in1}\exp(-k_2x\mathrm{i})+R_1^2\gamma\frac{T}{1-R^2\eta}A_{in1}\exp(-k_2x\mathrm{i})+R_1^4\gamma^2\frac{T}{1-R^2\eta}A_{in1}\exp(-k_2x\mathrm{i})+\cdots$$
$$=\frac{T}{(1-R^2\eta)(1-R_1^2\gamma)}A_{in1}\exp(-k_2x\mathrm{i})=B_{n1}^{(0)}\exp(-k_2x\mathrm{i}), \tag{7}$$

$$w_{nb-}^{(0)}(x)=R_1\tau_1\eta_1\frac{T}{1-R^2\eta}A_{in}\exp(k_2x\mathrm{i})+R_1^3\gamma\tau_1\eta_1\frac{T}{1-R^2\eta}A_{in}\exp(k_2x\mathrm{i})+\cdots$$
$$=R_1\tau_1\eta_1\frac{T}{(1-R^2\eta)(1-R_1^2\gamma)}A_{in}\exp(k_2x\mathrm{i})=B_{n2}^{(0)}\exp(k_2x\mathrm{i}), \tag{8}$$

where $\tau_1=\exp(-2k_2l\mathrm{i})$, $\eta_1=\exp(-2k_2a\mathrm{i})$, $\gamma=\exp(-2k_2a_2\mathrm{i})$, R_1 is the reflection coefficient of the segment (1).

Similarly, if waves are transmitted in only leftward direction, then wave-field established on segment (0) $w_{na-}^{(0)}(x)$ will be transmitted past discontinuity (0), and the total transmission will suffer multiple reflections between discontinuity (-1) and (0). Multiple reflections produce a total wave $w_{(n-1)b-}^{(0)}(x)$ and a total wave $w_{(n-1)b+}^{(0)}(x)$ on segment (-1) by superposition, where

$$w_{(n-1)b-}^{(0)}(x)=\frac{TR\tau\eta}{(1-R^2\eta)(1-R_1^2\gamma)}A_{\mathrm{in}1}\exp(k_2x\mathrm{i})=B_{(n-1)2}^{(0)}\exp(k_2x\mathrm{i}), \tag{9}$$

$$w_{(n-1)b+}^{(0)}(x)=R_1\tau_1^{-1}\gamma\frac{TR\tau\eta}{(1-R^2\eta)(1-R_1^2\gamma)}A_{\mathrm{in}1}\exp(-k_2x\mathrm{i})=B_{(n-1)1}^{(0)}\exp(-k_2x\mathrm{i}). \tag{10}$$

So wave-field on any segment (n) can also be established.

2.3 Bidirectional propagation

Here wave propagation in both directions is considered by means of an iteration procedure. In this procedure, each step deals with propagation in only one direction and gives a correction term that, in essence, accounts for the propagation direction not considered in the proceeding step [5]. Above Section give initial approximations that neglect the effects of propagation in the leftward direction on each segment when waves are transmitted in only rightward direction. The problem is taken into account subsequently.

The first correction terms on segment (0) are due to the leftward propagating components on all segments to the right of the segment (0) when only leftward propagation is permitted. $w_{nb-}^{(0)}(x)$ on segment (1) will be transmitted past discontinuity (1), and the total transmission will suffer multiple reflections between discontinuity (0) and (1). Multiple reflections produce a total wave on segment (0); $w_{(n+1)a-}^{(0)}(x)$ on segment (2) will be transmitted past discontinuity (2), and the total transmission will suffer multiple reflections between discontinuity (1) and (2), then will be transmitted past discontinuity (1), and the total transmission will suffer multiple reflections between discontinuity (0) and (1). Multiple reflections produce a total wave on segment (0). By the same token, the first correction terms $w_{na-}^{(1)}(x)$ and $w_{na+}^{(1)}(x)$ can be obtained by superposing the waves resulting on segment (0)

$$w_{na-}^{(1)}(x)=\left(PB_{n2}^{(0)}+PP_1A_{(n+1)2}^{(0)}+P^2P_1B_{(n+1)2}^{(0)}+\cdots\right)\exp(k_1x\mathrm{i}), \tag{11}$$

$$w_{na+}^{(1)}(x)=\tau^{-1}R\left(PB_{n2}^{(0)}+PP_1A_{(n+1)2}^{(0)}+P^2P_1B_{(n+1)2}^{(0)}+\cdots\right)\exp(-k_1x\mathrm{i}), \tag{12}$$

where $T_1/(1-R^2\eta)=P$, $T/(1-R_1^2\gamma)=P_1$, number of term in series equals to the number of segment to the right of the segment (0). The first correction terms on any segment can also be obtained in the same way.

Similar to the above considered case, the second correction terms on a given segment can be obtained. The second correction terms on segment (0) are due to the rightward propagating components on all segments to the left of the segment (0). $w_{na+}^{(2)}(x)$ and $w_{na-}^{(2)}(x)$ can be obtained by superposing the waves resulting on segment (0).

$$w_{na+}^{(2)}(x)=\left(PB_{(n-1)1}^{(1)}+PP_1A_{(n-1)1}^{(1)}+P^2P_1B_{(n-2)1}^{(1)}+\cdots\right)\exp(-k_1x\mathrm{i}), \tag{13}$$

$$w_{na-}^{(2)}(x)=\tau\eta R\left(PB_{(n-1)1}^{(1)}+PP_1A_{(n-1)1}^{(1)}+P^2P_1B_{(n-2)1}^{(1)}+\cdots\right)\exp(k_1x\mathrm{i}), \tag{14}$$

where number of term in series equals to the number of segment to the left of the segment (0). The second correction terms on any segment can also be obtained in the same way.
The third and the fourth correction terms can also be obtained by means of this iteration procedure. The third correction terms on a given segment are due to the leftward propagating components of the second correction terms on all segments to the right of the segment when only leftward propagation is permitted. The fourth correction terms on a given segment are due to the rightward propagating components of the third correction terms on all segments to the left of the segment. The wave-field on a given segment can be obtained by superposing the initial approximation and the various corrections. For example, the wave-field on segment (0) can be obtained

$$w_{na+}(x)=\sum_{i=0}^{\infty}w_{na+}^{(i)}(x)=W_{na+}\exp(-k_1x\mathrm{i}),\qquad w_{na-}(x)=\sum_{i=0}^{\infty}w_{na-}^{(i)}(x), \tag{15a}$$

where number of term in series is determined by iteration times.

Similarly, we have

$$w_{(n-1)b+}(x)=\sum_{i=0}^{\infty}w_{(n-1)b+}^{(i)}(x)=W_{(n-1)b+}\exp(-k_2x\mathrm{i}),\qquad w_{(n-1)b-}(x)=\sum_{i=0}^{\infty}w_{(n-1)b-}^{(i)}(x), \tag{15b}$$

$$w_{nb+}(x)=\sum_{i=0}^{\infty}w_{nb+}^{(i)}(x)=W_{nb+}\exp(-k_2x\mathrm{i}),\qquad w_{nb-}(x)=\sum_{i=0}^{\infty}w_{nb-}^{(i)}(x). \tag{15c}$$

The wave-field on any segment can also be obtained. So the amplitudes of the wave-field resulting from the incident propagating wave can be determined.

3. The determination of Iteration times

Iteration times may be determined by calculating the ratio of the wave-field on adjacent cells in the next step. Considering Eq. (15), we have

$$\overline{W}_{nb+}=\exp(-k_2(l+a_1)\mathrm{i})W_{nb+},\qquad \overline{W}_{(n-1)b+}=\exp(-k_2(l-a+a_1)\mathrm{i})W_{(n-1)b+}, \tag{16}$$

where the symbols with the bar denote components referred the local coordinate systems originating at the left ends of the segments. The Bloch theorem states that $\overline{W}_{nb+}=\exp(\mathrm{i}\mu a)\overline{W}_{(n-1)b+}$. So we have the following relation of the wave-field on adjacent cells

$$W_{nb+}=\exp(\mathrm{i}\mu a)\exp(k_2a\mathrm{i})W_{(n-1)b+}. \tag{17a}$$

Similarly, we can get

$$W_{na+}=\exp(\mathrm{i}\mu a)\exp(k_1a\mathrm{i})W_{(n-1)a+}. \tag{17b}$$

Comparing the propagating constant μ calculated by Eq. (17) with μ calculated by the transfer matrix method, iteration times of Eq. (15) can be determined within the desired accuracy [5].

4. Numerical examples and discussions

From this section, numerical calculations of band structures for finite periodic beams with epoxy as material A and aluminum as material B are performed. Material constants used in the calculation are mainly listed in Table 1. The lattice constant is chosen to be $a = 2\text{m}$, and $a_1 = a_2 = 1\text{m}$. The beam cross-sectional area is $A = 5.969\times10^{-3}\,\text{m}^2$ and the second area moment of inertia is $I = 2.701\times10^{-5}\,\text{m}^4$ for segments A and B . The distance between the origin of the x system and the discontinuity (0) is $l = 3a$.

Table 1
Material constants

Materials	Yong's modulus E (Gpa)	Density ρ (kg/m^3)
Epoxy (A)	4.35	1180
Aluminum (B)	77.56	2730

4.1 Band gap of periodic beam

The band structure of an infinite periodic beam is depicted in Fig. 2. The shadow areas are the first two band gaps.. It is repeated here for the sake of completeness and comparison.

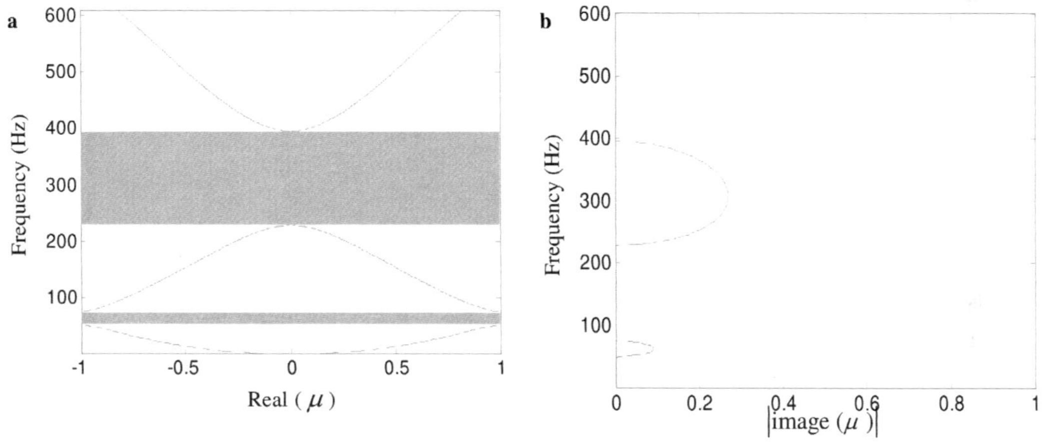

Fig. 2 Band structure (dispersion relationship) for an infinite periodic beam. (a) Real wave vector. (b) The absolute value of the imaginary part of complex wave vector (propagating constant).

4.2 Multi-reflection approach

4.2.1 The number of cells In this section, the frequency response on different segment of the beam is given to describe the band gap characteristics of finite periodic beams. In Fig. 3, Frequency response functions curves of finite periodic beams considering the propagating disturbance are given. Curves are shown for M=6, 10 and 15 (the number of cell).

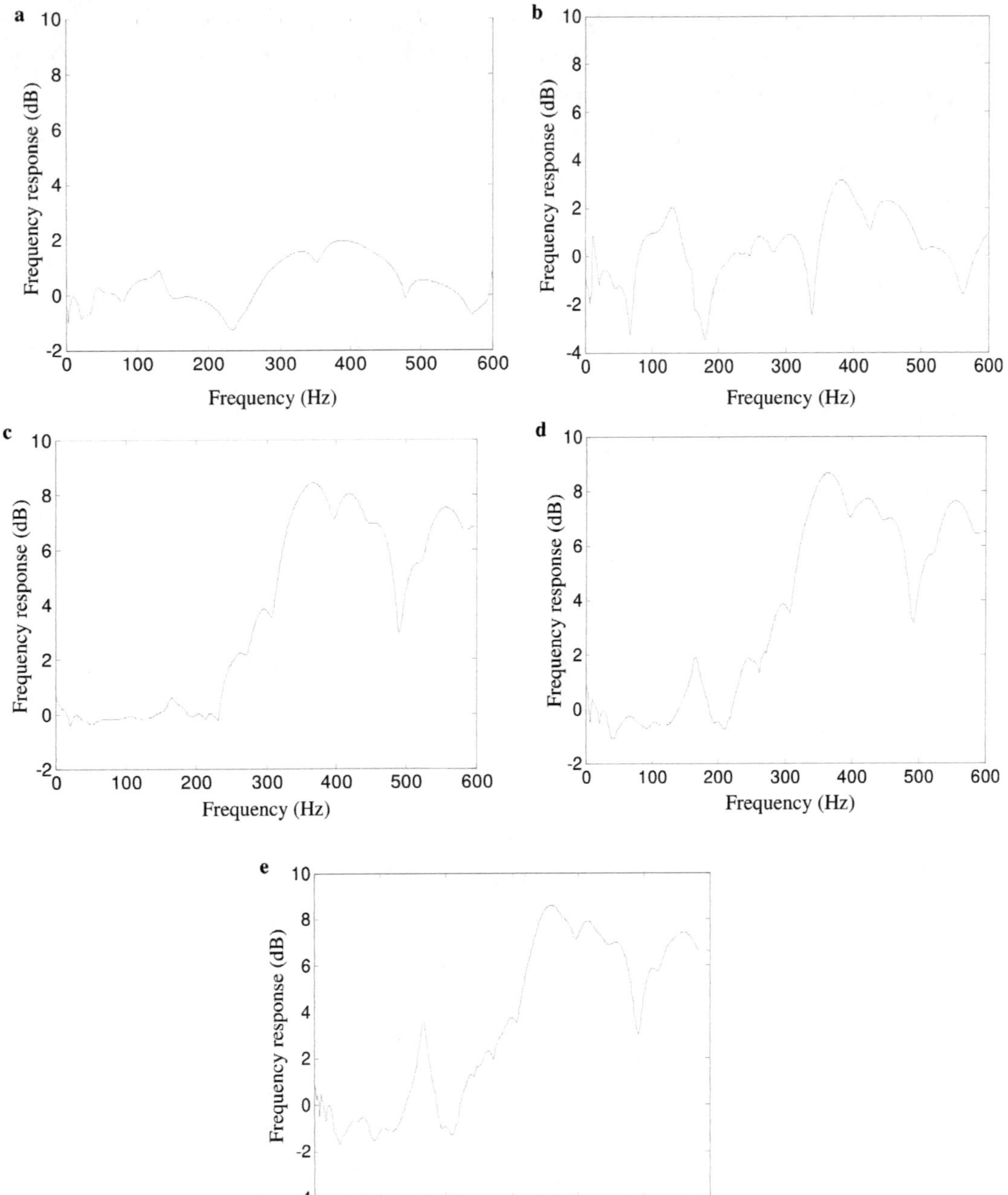

Fig. 3 Frequency response of periodic beams. (a) M=6, segment (4), Iteration times N=5, (b) M=10, segment (12), N=5, (c) M=15, segment (4), N=9, (d) M=15, segment (12), N=9, (e) M=15, segment (22), N=9.

In Fig. 3a, 3b and 3e, we can see obviously that the band gaps of a finite periodic beam is more agreement with the band gaps of infinite periodic beam for larger M . However, the first band gap is not obvious. We can also observe from Fig. 3c, 3d and 3e that the longer the distance from the segment (0) (propagating disturbance), the more obvious the band gaps become. It is shown from Fig. 3a and 3c that the band gap characteristics can not be found because the distance from the segment (0) (propagating disturbance) is too short. So the longer the distance from disturbance, the more obvious the band gaps become for periodic structures with a lateral disturbance force.

4.2.2 Effect of the material parameters In what follows, parametric studies are undertaken to investigate the effect of the ratio ρ_A / ρ_B and E_A / E_B on the band gaps in periodic beams with disturbance. Frequency response on segment (22) of periodic beams is considered. M=15, N=9. The material A is assumed to be epoxy, however, the density of material B varies with the change of ratio ρ_A / ρ_B. All other parameters are taken the same as above. The effect of ρ_A / ρ_B on the second band gap is plotted in Fig. 4. The bandwidth of the first band gap is too small to consider. The beginning frequency of the second band gap increases and the cutoff frequency decreases with the increasing of the ratio ρ_A / ρ_B. Now the Young's modulus of the beam is considered. We consider the case in which the material A is epoxy. The density of material B varies with the change of ratio E_A / E_B. All other parameters are taken the same as above. The effect of E_A / E_B on the second band gap is plotted in Fig. 5. It can be observed from Fig. 5 that both the beginning frequency and cutoff frequency of the second band gap are decrease with the increasing of the ratio E_A / E_B.

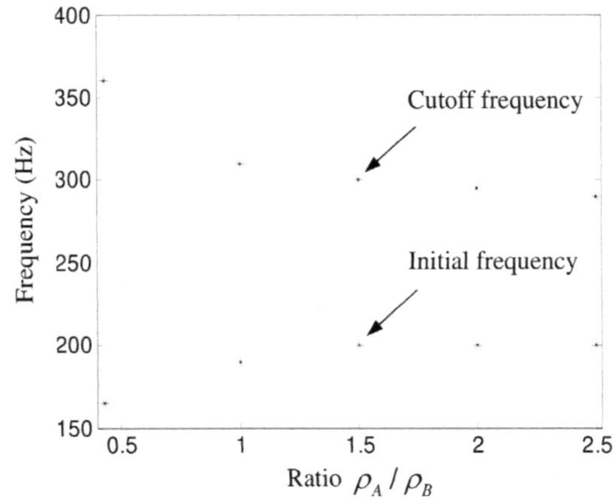

Fig. 4 Effect of the ratio ρ_A / ρ_B on the second band gap.

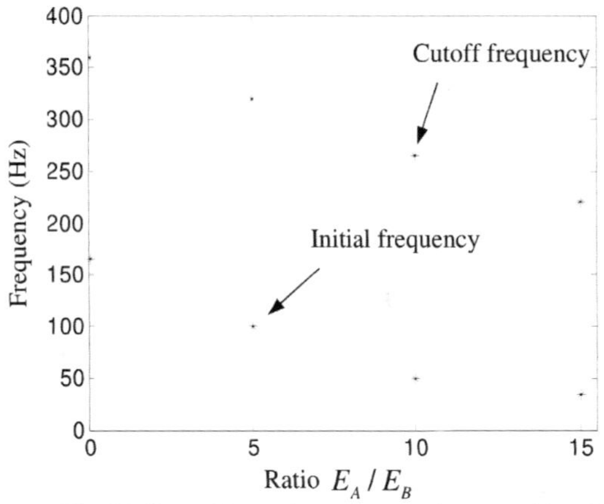

Fig. 5 Effect of the ratio E_A / E_B on the second band gap.

4.2.3 Effect of the disturbance Fig. 6 shows the frequency responses for the periodic beams with different the propagating disturbance. The material constants used in the calculation are the same as those in Table 1. Although the band gap characteristics can not be changed by tuning the disturbance, it can be seen that the attenuation of the band gaps gradually decreases with the increase disturbance amplitude.

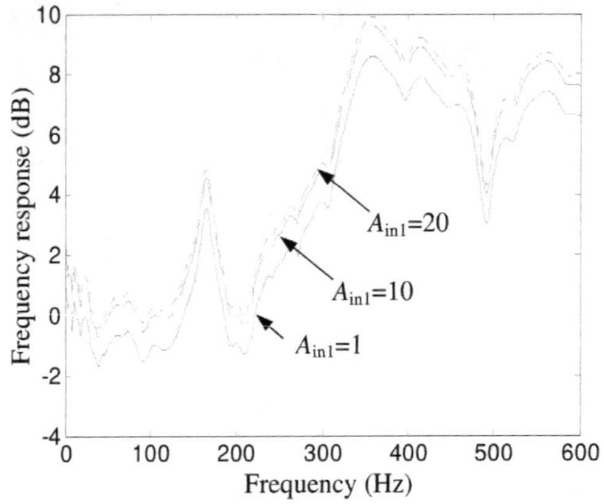

Fig. 6 Frequency response of periodic beams with different propagating disturbance. M=15, segment (22), N=9.

5. Conclusions

The elastic wave propagation in finite periodic beams considering the propagating disturbance is studied theoretically as well as numerically. The effects of the disturbance on the wave propagation in periodic beams are considered. Multi-reflection method is used to analyze the band gap characteristics of the periodic beam with one or more lateral disturbance. It is very difficult to study effects of the disturbance on the wave propagation in periodic beams using transfer matrix method. The validity of the multi-reflection is proved. The amplitudes of the wave-field on any segment resulting from a given incident amplitude can be determined. It is also not given by transfer matrix method. Expression is presented from which one may evaluate the propagation constant, and then study wave propagation characteristics of a finite periodic beam considering a propagating disturbance. Much attention is devoted to the response in the frequency ranges with gaps in the band structure for the corresponding periodic beams. From the results, the following conclusions can be drawn:

(1) The effects of the disturbance on the wave propagation in periodic beams is analyzed, and a relation between the amplitude of the incident wave and the resulting amplitudes of any segment is established using multi-reflection method, and these can not be given using the transfer matrix method.

(2) For the finite periodic beam, band gap characteristic are not obvious when the distance between the segment and the disturbance is short. It is shown that the influence of disturbance on the band gaps of the periodic beam is focus on the areas in the vicinity. Band gap characteristic will disappear because of the influence of external disturbance. At same time, the longer the distance from the disturbance, the more obvious the band gaps become, and the band gaps of the finite periodic beam is more agreement with the band gaps of the infinite periodic beam for larger M . The validity of the multi-reflection is proved.

(3) Frequency response of finite periodic beams can be changed by tuning the disturbance. With the increase disturbance amplitude, the attenuation of the band gaps gradually decreases.

Acknowledgements

The work was supported by the National Natural Science Foundation of China under Grant No. 11102047,11002037 and the Fundamental Research Foundation of Harbin Engineering University under Grant No. 002110260746.

References

[1] Brillouin L 1953 New York: Dover.

[2] Heckl M J 1964 Acoust Soc Am 36 1335-1343.
[3] Mead D J 1971 ASME J Engi Indust 21 783-792.
[4] Mead D J 1996 J Sound Vib190 495-524.
[5] Ungar E E 1965 J Acoust Soc Am 39 887-894.
[6] Sigalas M M and Economou EN 1993 Solid State Commun 86 141-143.
[7] Kushwaha M S etal 1993 Phys Rev Lett 71 2022-2025.
[8] Tanaka Y and Tamura S 1998 Phys Rev B 58 7958-7965.
[9] Ruzzene M and Baz A 2000 ASME J Vibr Acoust 122 151-159.
[10] Vasseur J O etal 2001 Phys Rev Lett 86 3012-3015.
[11] Liu Z Y etal 2005 Phys Rev B 71 014103.
[12] Zhang X Y etal 2006 Appl Phys Lett 88 041911.
[13] Wang Y Z etal 2009 Acta Mech Sinica 25 65-71.
[14] Golub M V etal 2012 Int J Solids Struct 49 344-354.
[15] Jensen J S 2003 J Sound Vib 266 1053-1078.
[16] Kobayashi F etal 2004 Int J Solids Struct 41 7361-7375.
[17] Xiang H J and Shi Z F 2009 Comp and Struct 87 1559-1566.
[18] Yu D L etal 2011 Phys Lett A 376 626-630.

Journal of Physics: Conference Series **448** (2013) 012013 doi:10.1088/1742-6596/448/1/012013

In-plane motion of wind turbine blade under the combined action of parametric and forced excitations

L Li[1], Y H Li[1,3], H W Lv[1] and E C Yang[1,2]
[1]School of Mechanics and Engineering, Southwest Jiaotong University, Chengdu 610031, China
[2]Chongqing Automobile Institute, Chongqing University of Technology, Chongqing 400054, China

Email: yinghui.li@home.swjtu.edu.cn

Abstract. This paper presents an analysis for the nonlinear in-plane motion of wind turbine blade under the combined action of parametric and forced excitations. The Bernoulli-Euler beam theory is adopted to describe blade motion. Effects including gravity, structural damping and geometric nonlinearity are included in the mathematical model. The rotating speed is considered as a nominal speed with a harmonic perturbation. The periodic oscillation of rotating speed brings about parametric excitations in stiffness and damping terms and forced excitations. The harmonic balance method is applied to get the approximate solution of dynamic response. Numerical integral of original system is used to verify the analytical solution. Influences of structural damping, perturbed amplitude and frequency of rotating speed on blade behaviour are discussed.

1. Introduction

Hub motion can cause the oscillation of rotating speed that will bring about parametric and forced excitations for a wind turbine blade. In studying parametric excitation of wind turbine blade, Chopra and Dugundji [1] studied principal parametric resonance and self-excited flutter of the coupled feather-flap-lead/lag vibration, while the blade was considered as a rigid body with hinges; effects of design parameters on dynamic responses were analyzed. Larsen et al. [2−4] adopted Bernoulli-Euler beam theory to describe the coupled vibration of slender blade and investigated nonlinear response, parametric instability and stochastic stability of blade in 1:2 internal resonance associated with parametric excitation; influences of aerodynamic damping, perturbed amplitude, external excitation, linear and nonlinear parametric terms on dynamic responses were discussed. Ramakrishnan and Feeny [5] introduced a mathematical model for blade in-plane motion with parametric excitation caused by gravitational load by employing extended Hamiltonian principle, and presented an analysis for dynamic responses of blade in super-harmonic resonance; effects of ratio between rotating angular frequency and natural frequency, external excitation, parametric forcing term and nonlinearity on blade behaviours were studied. Except these literature for wind turbine blades, there are lots of works on parametric excitations, e.g. [6−17], for beams, plates, strings, cables and rigid bodies.

Previous researches are mainly on parametric excitations in resonance cases. In this work, the nonlinear in-plane motion (the most unstable motion for a blade due to small aerodynamic and

[3] Corresponding author: Y H Li.

structural damping) of wind turbine blade under the combined actions of parametric and forced excitations in general case is studied. The structural damping which has been studied scarcely in previous researches about nonlinear dynamics of blade is added to the model via the Kelvin-Voigt principle. The harmonic balance method is adopted to obtain the approximate solution of dynamic responses. Influences of structural damping, perturbed frequency and perturbed amplitude on blade behaviours are discussed.

2. Mathematical model

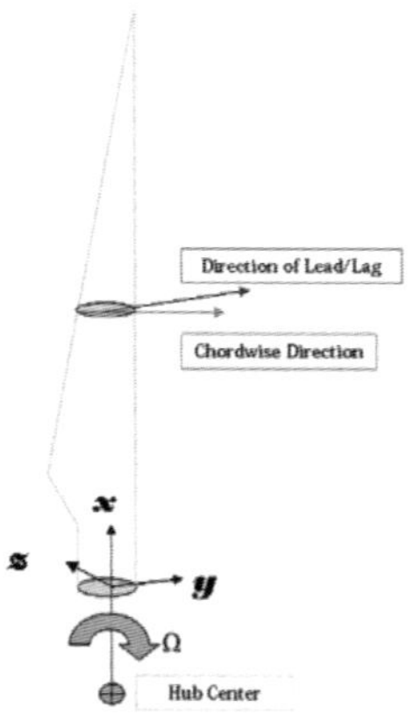

Figure 1. Static configuration of an isolated blade.

A sketch of an isolated blade is illustrated in Figure 1. The in-plane bend is called lead/lag. The Cartesian coordinate system (xyz) is a body coordinate system which is rigidly attached to the blade root such that the x-axis corresponds to the undeformed elastic axis, the y-axis is along the lead/lag direction, and the z-axis is perpendicular to the plane of rotation. The Euler-Bernoulli beam theory is adopted to describe the vibration of slender blade driven by wind flow. The structural damping is introduced to the model through the Kelvin-Voigt principle ($\sigma = E\varepsilon + \bar{\eta}\dot{\varepsilon}$); a detailed deduction process of structural damping terms is given in [18] and [19]. The governing equation is derived as:

$$\left(EIv''\right)'' - \left(F_T v'\right)' - m\Omega^2\left(e_y + v\right) + m\ddot{v} - \left(\bar{\eta}Av'^2\dot{v}'\right)' + \left(\bar{\eta}I\dot{v}''\right)'' = F_y \tag{1}$$

Where E is the elastic modulus, I is the area moment inertia, v is the bending displacement, $F_T = \int_A \sigma \mathrm{d}A = EA(v'^2/2)$ is the axial force whose steady part $N_p = \int_x^L m(y)\Omega^2 (e + y\cos\beta_p)\mathrm{d}y$ is the centrifugal force, m is the mass per unit length, Ω is the rotating angular speed, (e_x, e_y) is the blade root offset from hub center in (x, y)-direction due to inclined installation, $\bar{\eta}$ is the viscosity factor, A is the section area, F_y is the resultant distributed force in y-direction. Gravity and aerodynamic force are considered as distributed forces acting on blade sections. Gravity has the following form:

$$F_y^{(G)}(x) = -mg\sin\left(\Omega t\right) \tag{2}$$

The following unsteady aerodynamic force given in Li et al. [20] that is derived from Greenberg's expressions is adopted in this paper:

$$\begin{aligned} F_y^{(A)} = {} & \left(\rho ac/2\right)\Omega^2\left[\lambda^2 R^2\cos\theta - \lambda\left(1-\beta_p^2\right)Rx\sin\theta - \lambda Re_x\sin\theta - \left(C_{DO}/a\right)x^2\right] \\ & -\left(\rho ac/2\right)\left\{\Omega\left[\lambda R\sin\theta + \left(2C_{DO}/a\right)x\right]\dot{v} + \left(c/4\right)\ddot{v}\sin^2\theta\right\} \end{aligned} \tag{3}$$

where ρ is the air density, a is the section lift curve slope, c is the chord length, $\lambda=\lambda_0+\lambda_1(r_x/R)^2+\lambda_2\cdot(r_x/R)\cos(\Omega t)$ is the rotor inflow ratio ($1/\lambda$ is the blade tip speed ratio), λ_0 is the inflow ratio at hub, $\lambda_1=[p(p-1)/4](R/h_0)^2(\lambda_0+\bar{\sigma}a/8)$, p is the velocity gradient constant, R is the rotor radius, h_0 is the hub height from ground level , $\bar{\sigma}=Nc/(\pi R)$ is the solid ratio, N is the blade number in a wind turbine, $r_x=(x+e_x)\cos\beta_p$ is the blade radial station from root, $\lambda_2=(-pR/h_0)(\lambda_0+\bar{\sigma}a/8)$, $\theta=\theta_0+\theta_t(x)$ is the twist angle from chordwise direction to lead/lag direction, θ_0 is the setting angle, θ_t is the pre-twist angle, β_p is the coning angle between blade and vertical plane, C_{DO} is the drag coefficient.

Due to hub motion, there are oscillations for the nominal speed Ω_0 of a rotor blade in operating process. The practical speed can be expressed in Fourier-series:

$$\Omega(t)=\Omega_0+\sum_{n=1}^{\infty}a_n\cos(\Omega_n t)+\sum_{n=1}^{\infty}b_n\sin(\bar{\Omega}_n t) \tag{4}$$

To discussed the effects of speed oscillation more distinctly, the speed Ω is denoted as

$$\Omega=\Omega_0+\varepsilon\cos(\Omega_p t+\varphi_p) \tag{5}$$

where the perturbed amplitude ε is a small amount of Ω_0. Galerkin's projection is employed to discrete the continuous model. The displacement is expressed as

$$v=\gamma(x)q(t) \tag{6}$$

where γ (x) is the first-order undamped modal function; it is obtained by using the numerical integral basing on Green function [21]. $q(t)$ is the modal coordinate that satisfies the following equation:

$$\begin{aligned}&\ddot{q}+\omega^2 q-2\hat{m}\varepsilon\Omega_0\cos(\Omega_p t+\varphi_p)q+\Omega c_a\dot{q}+c_{s1}\dot{q}+c_{s2}\dot{q}q^2\\&-\Omega^2 a_s+(bq^3-\Omega_0^2\hat{b}q)=\Omega^2 f\cos(\Omega_0 t)+\hat{g}\sin(\Omega_0 t)\end{aligned} \tag{7}$$

Where ω is the linear frequency, the third term is the time-dependent elastic restoring force, c_a-term is the aerodynamic damping force, c_{s1}- and c_{s2}-terms are structural damping forces, a_s-term is the linear static displacement, the last term preceding equals sign is the dynamic part of axial force, f-term is the aerodynamic excitation, the last term is gravity. Expressions of coefficients are given in Appendix A.

The generalized displacement is disintegrated into static and dynamic parts: $q(t)=q_s+q_d(t)$. Equations for two parts are obtained as:

$$\omega^2 q_s-\Omega_0^2 a_s+bq_s^3-\Omega_0^2\hat{b}q_s=0 \tag{8}$$

$$\begin{aligned}&\ddot{q}_d+\hat{\omega}^2 q_d-2\hat{m}\varepsilon\Omega_0\cos(\Omega_p t+\varphi_p)(q_s+q_d)+\left[\Omega_0+\varepsilon\cos(\Omega_p t+\varphi_p)\right]c_a\dot{q}_d\\&+c_{s1}\dot{q}_d+c_{s2}\dot{q}_d(q_s+q_d)^2-2\varepsilon\Omega_0\cos(\Omega_p t+\varphi_p)a_s+b(3q_s q_d^2+q_d^3)\\&=\left[\Omega_0+\varepsilon\cos(\Omega_p t+\varphi_p)\right]^2 f\cos(\Omega_0 t)+\hat{g}\sin(\Omega_0 t)\end{aligned} \tag{9}$$

where $\hat{\omega}=\sqrt{\omega^2+3bq_s^2}$ is the nonlinear frequency due to geometric nonlinearity. In equation (9), $(2\pi/\Omega_p)$-periodic stiffness, aerodynamic damping and external excitations appear due to rotating speed oscillation. The equation of static displacement can be solved by numerical method directly. The equation of dynamic displacement is solved by the harmonic balance method in this paper.

3. Analytical solution

$(2\pi/\Omega_0)$- and $(2\pi/\Omega_p)$-periodic terms in dynamic response are emphasized by neglecting small higher-order-harmonic-terms. The solution is assumed to be:

$$q_d \approx a_f \sin(\Omega_0 t) + b_f \cos(\Omega_0 t) + c_f \sin(\Omega_p t) + d_f \cos(\Omega_p t) \tag{10}$$

Substituting equations (5) and (10) into equation (9) and equating the same harmonic terms, the following algebraic equations are obtained:

$$\begin{aligned}&\left(\hat{\omega}^2 - \Omega_0^2\right)a_f - \left(\Omega_0^2 c_a + \Omega_0 c_{s1} + \Omega_0 c_{s2} q_s^2\right)b_f - \frac{1}{4}\Omega_0 c_{s2} b_f \left(a_f^2 + b_f^2 + 2c_f^2 + 2d_f^2\right)\\&+\frac{3}{4} b a_f \left(a_f^2 + b_f^2 + 2c_f^2 + 2d_f^2\right) = \hat{g}\end{aligned} \tag{11}$$

$$\begin{aligned}&\left(\Omega_0^2 c_a + \Omega_0 c_{s1} + \Omega_0 c_{s2} q_s^2\right)a_f + \left(\hat{\omega}^2 - \Omega_0^2\right)b_f + \frac{1}{4}\Omega_0 c_{s2} a_f \left(a_f^2 + b_f^2 + 2c_f^2 + 2d_f^2\right)\\&+\frac{3}{4} b b_f \left(a_f^2 + b_f^2 + 2c_f^2 + 2d_f^2\right) = \Omega_0^2 f\end{aligned} \tag{12}$$

$$\begin{aligned}&\left(\hat{\omega}^2 - \Omega_p^2\right)c_f - \left(\Omega_0 \Omega_p c_a + \Omega_p c_{s1} + \Omega_p c_{s2} q_s^2\right)d_f - \frac{1}{4}\Omega_p c_{s2} d_f \left(2a_f^2 + 2b_f^2 + c_f^2 + d_f^2\right)\\&+\frac{3}{4} b c_f \left(2a_f^2 + 2b_f^2 + c_f^2 + d_f^2\right) = -2\hat{m}\varepsilon\Omega_0 q_s \sin\varphi_p - 2\varepsilon\Omega_0 a_s \sin\varphi_p\end{aligned} \tag{13}$$

$$\begin{aligned}&\left(\Omega_0 \Omega_p c_a + \Omega_p c_{s1} + \Omega_p c_{s2} q_s^2\right)c_f + \left(\hat{\omega}^2 - \Omega_p^2\right)d_f + \frac{1}{4}\Omega_p c_{s2} c_f \left(2a_f^2 + 2b_f^2 + c_f^2 + d_f^2\right)\\&+\frac{3}{4} b d_f \left(2a_f^2 + 2b_f^2 + c_f^2 + d_f^2\right) = 2\hat{m}\varepsilon\Omega_0 q_s \cos\varphi_p + 2\varepsilon\Omega_0 a_s \cos\varphi_p\end{aligned} \tag{14}$$

Four harmonic amplitudes can be derived from equations (11)–(14) through a numerical computation.

4. Numerical results and discussion

The data for a three-bladed MW wind turbine are taken in analysis. Blade length L=48m, hinge offset e_x=e_y=0.035L, elastic modulus E=3×10^{10} Pa, blade density ρ_b=1.8×10^3 kg/m^3, nominal speed Ω_0=25r/min, section lift curve slope a=2π, drag coefficient C_{DO}=0.018, hub height h_0=2L, air density ρ=1.25kg/m^3, velocity gradient constant p=0.167, gravitational acceleration g=9.8m/s^2.

Changes of amplitudes with structural damping ratio $\zeta = c_{s1}/(2\hat{\omega})$, perturbed amplitude ε and frequency ratio $\gamma = \Omega_p/\hat{\omega}$ for φ_p=0 are illustrated in figure 2. φ_p=0 represents the cosine oscillation Ω=Ω_0+εcos(Ω_pt). It shows that four amplitudes all tend to constants with increasing γ, especially that c_f and d_f approach 0 for large γ; this implies that the steady state response is a (2π/Ω_0)-periodic motion for large γ. There's a noticeable change near γ=1 for each amplitude due to the primary resonance arising from (2π/Ω_p)-harmonic forced excitations; this change becomes dramatic with increasing ±ε and decreasing ζ. For small γ, the harmonic amplitude c_f is near zero; so the (2π/Ω_p)-periodic term is mainly cos(Ω_pt)-term arising from forced excitations, and the absolute value of the amplitude d_f of the (2π/Ω_p)-periodic term increases with increasing ±ε and decreasing ζ. ε and −ε generate same amplitudes.

To reveal the effects of the sine oscillation Ω=Ω_0+εsin(Ω_pt), φ_p=−π/2 is taken in analysis and curves of four harmonic amplitudes with respect to ε, γ and ζ are presented in figure 3. It is the same with conclusions for cosine oscillation that (2π/Ω_p)-periodic terms are very tiny for large γ, and there is a obvious change for each amplitude near γ=1 that increases with increasing ±ε and decreasing ζ. For small γ, the harmonic amplitude d_f is almost zero; so the (2π/Ω_p)-periodic term is mainly sin(Ω_pt)-term arising from forced excitations, and the absolute value of the amplitude c_f of the (2π/Ω_p)-periodic term increases with increasing ±ε and decreasing ζ. ε and −ε generate same amplitudes.

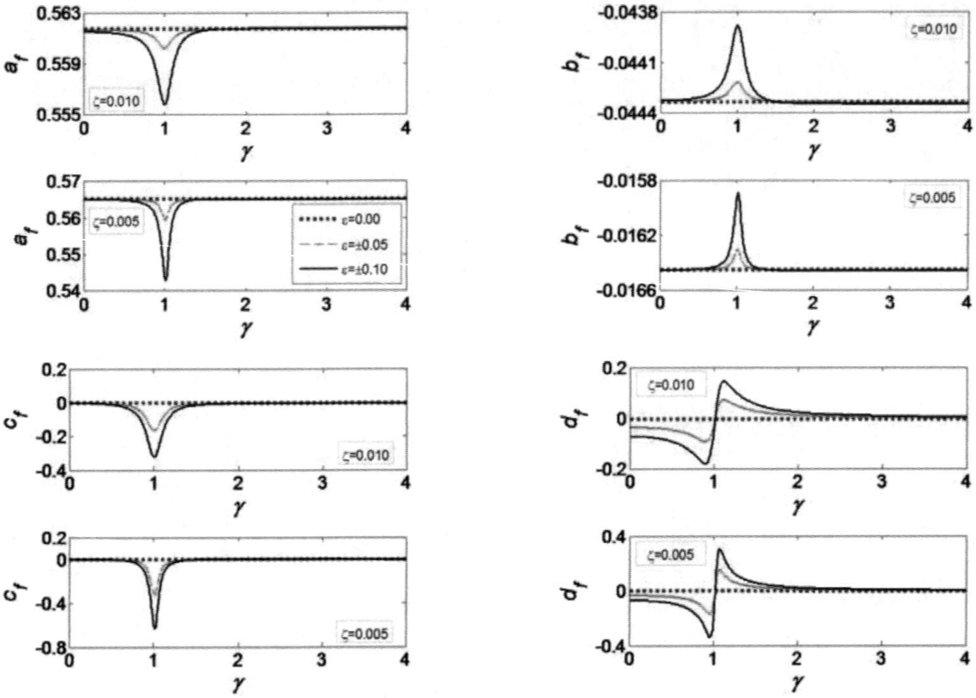

Figure 2. Variations of four harmonic amplitudes with respect to structural damping ratio ζ, perturbed amplitude ε and frequency ratio γ, where φ_p=0.

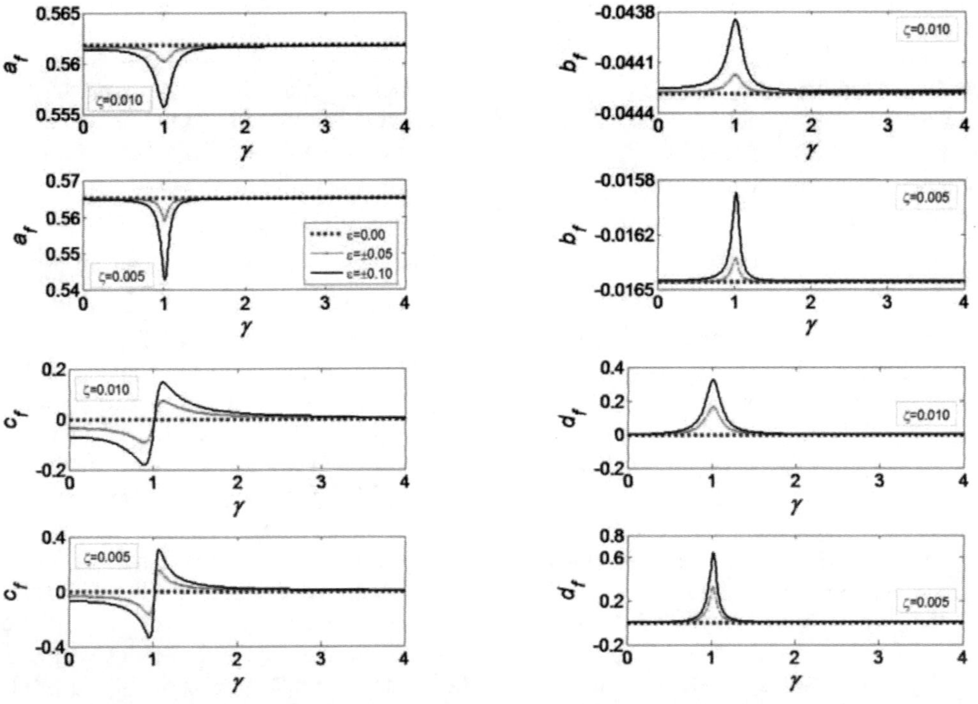

Figure 3. Variations of four harmonic amplitudes with respect to structural damping ratio ζ,

perturbed amplitude ε and frequency ratio γ, where $\varphi_p=-\pi/2$.

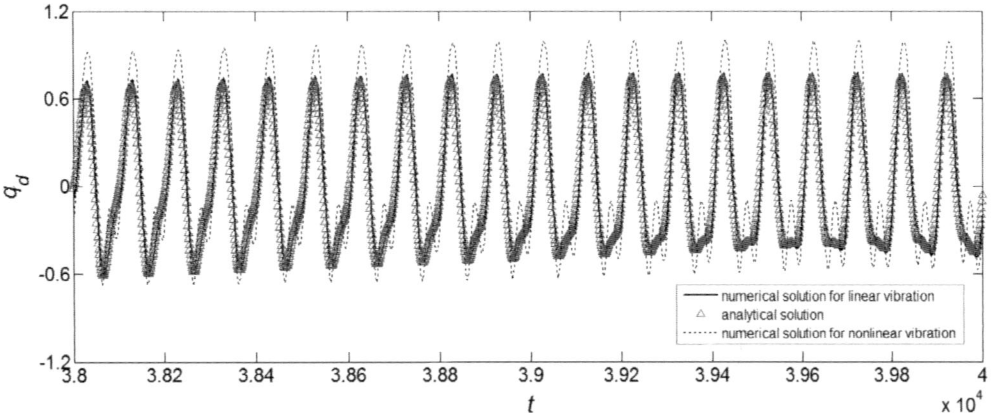

Figure 4. Time history of dynamic displacement for ε=0.10, ζ=0.005, γ=0.8, φ_p=0.

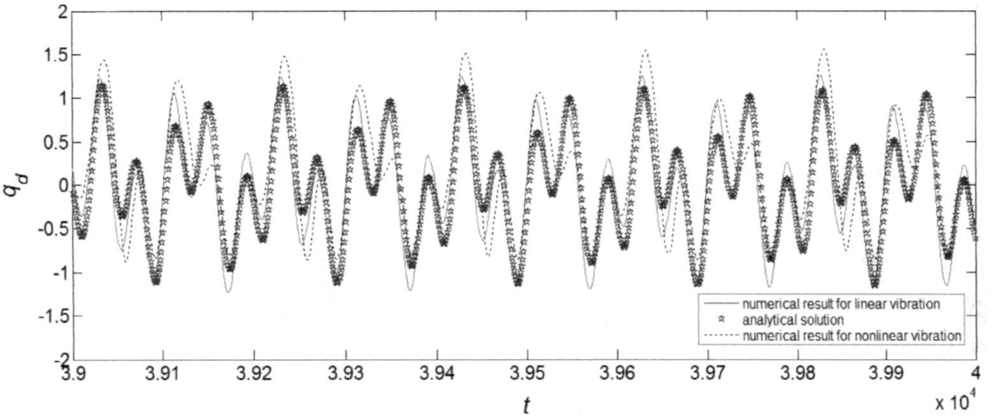

Figure 5. Time history of dynamic displacement for ε=0.10, ζ=0.005, γ=1, $\varphi_p=-\pi/2$.

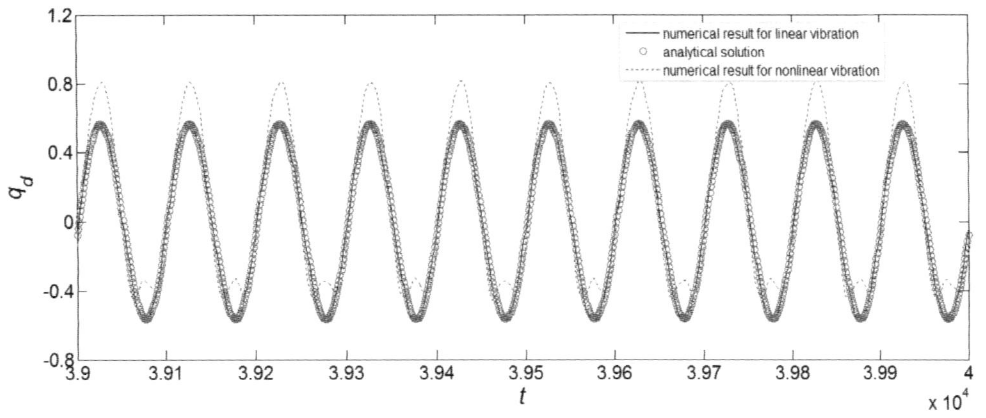

Figure 6. Time history of dynamic displacement for ε=0.10, ζ=0.005, γ=3, φ_p=0.

To validate analytical solutions, time histories of the dynamic displacement for three different frequency ratios, γ=0.8, 1 and 3, are shown in figures 4–6. The numerical results are derived from the direct numerical integral of the system (9). Figure 4 illustrates the dynamic response of the in-plane motion for ε=0.10, ζ=0.005, γ=0.8 and φ_p=0. γ=0.8 represents the case that the vibration is near the primary resonance for $(2\pi/\Omega_p)$-periodic forced excitations. The time interval in this figure is $2\times(2\pi/\Omega_0)$. One finds that the analytical solution coincides with the numerical result for linear vibration very well. There is a difference between analytical solution and numerical result of nonlinear vibration because of the nonlinear effects. The dynamic response is not a periodic motion obviously.

The primary resonant response (γ=1) is given in figure 5. The parameters are taken to be ε=0.10, ζ=0.005 and $\varphi_p=-\pi/2$. The time interval is $(2\pi/\Omega_0)$. It shows that analytical solution and two numerical results don't coincide with each other well. The periodic property of analytical solution is the same with that of linear vibration. However, due to nonlinearities, the periodic property and amplitude of analytical result do not coincide with those of nonlinear vibration very well.

Dynamic response for a large perturbed frequency is presented in figure 6. The time interval is taken to be $(2\pi/\Omega_0)$. The parameters are selected as ε=0.10, ζ=0.005, γ=3 and φ_p=0. One finds that the analytical solution coincides with linear vibration very well. However, there is a difference in amplitudes between the analytical result and nonlinear vibration that arises from nonlinear effects. The period of blade motion is $(2\pi/\Omega_0)$. So $(2\pi/\Omega_p)$-harmonic terms are very tiny in dynamic displacement for large perturbed frequency.

5. Conclusion

In this work, the nonlinear in-plane motion of a wind turbine blade under the combined actions of parametric and forced excitations was analyzed. The Bernoulli-Euler beam theory was adopted to describe the vibration of the slender blade. Gravity, structural damping and large deformation were included in mathematical model. The rotating speed was assumed to be a nominal speed with a cosine oscillation. The harmonic balance method was used to get the approximate solution of the dynamic response by selecting the fundamental harmonic terms of nominal rotating angular frequency and perturbed frequency. Influences of structural damping, perturbed frequency and perturbed amplitude on blade motion were discussed. The following conclusions were derived: (1) the harmonic oscillation of rotating speed brings about time-dependent stiffness, periodic-changing aerodynamic damping, and forced excitations; (2) for small perturbed frequency, the influence of rotating speed oscillation is mainly in generating forced excitations; harmonic amplitude increases with increasing perturbed amplitude and decreasing structural damping; (3) when perturbed frequency is near natural frequency, the speed oscillation can introduce the primary resonance; in this case harmonic amplitudes depend on structural damping and perturbed amplitude sensitively; (4) for large perturbed frequency, harmonic terms corresponding to rotating speed oscillation are very tiny; effects of structural damping and perturbed amplitude on blade motion are little.

Acknowledgments

Authors wishing to acknowledge financial supports from the National Natural Science Foundation of China (no: 11072204), the Fundamental Research Funds for the Central Universities (no: SWJTU11ZT15) and Shanghai Key Laboratory of Mechanics in Energy Engineering.

Appendix A. Coefficients of discrete model

$$d_1=\left[p(p-1)\bar{\sigma}a/32\right](r_x/h_0)^2 \quad d_2=-(p\bar{\sigma}a/8)(r_x/h_0)$$
$$d_3=1+\left[p(p-1)/4\right](r_x/h_0)^2 \quad d_4=-p(r_x/h_0) \tag{A.1}$$

$$\bar{m}=\int_0^L\left(m+\rho ac^2\sin^2\theta/8\right)\gamma^2\mathrm{d}x \qquad \hat{m}=\frac{1}{\bar{m}}\int_0^L m\gamma^2\mathrm{d}x \tag{A.2}$$

$$c_a = \frac{1}{\bar{m}}\int_0^L \rho acx\left(C_{DO}/a\right)\gamma^2 \mathrm{d}x + \frac{1}{\bar{m}}\int_0^L \left(\rho ac/2\right) R\sin\theta\left(d_1 + \lambda_0 d_3\right)\gamma^2 \mathrm{d}x$$
$$c_{s1} = \frac{\bar{\eta}}{\bar{m}}\int_0^L \left(I\gamma''\right)'' \gamma \mathrm{d}x \qquad c_{s2} = -\frac{\bar{\eta}}{\bar{m}}\int_0^L \left(A\gamma'^3\right)' \gamma \mathrm{d}x \tag{A.3}$$

$$b = -\frac{1}{2\bar{m}}\int_0^L \left(EA\gamma'^3\right)' \gamma \mathrm{d}x \quad \hat{b} = -\frac{1}{\bar{m}}\int_0^L \left[\gamma'\int_x^L m\left(e + y\cos\beta_p\right)\mathrm{d}y\right]' \gamma \mathrm{d}x \quad \hat{g} = -\frac{1}{\bar{m}}\int_0^L mg\gamma \mathrm{d}x \tag{A.4}$$

$$a_s = \frac{1}{\bar{m}}\int_0^L me_y\gamma \mathrm{d}x - \frac{1}{\bar{m}}\int_0^L \left(\rho ac/2\right)\left(C_{DO}/a\right)x^2\gamma \mathrm{d}x + \frac{1}{\bar{m}}\int_0^L \left(\rho ac/2\right)R^2\cos\theta\lambda_0^2\left(d_3^2 + \frac{d_4^2}{2}\right)\gamma \mathrm{d}x$$
$$-\frac{1}{\bar{m}}\int_0^L \left(\rho ac/2\right)\left[\left(1-\beta_p^2\right)Rx\sin\theta + Re_X\sin\theta\right]\left(d_1 + \lambda_0 d_3\right)\gamma \mathrm{d}x \tag{A.5}$$
$$+\frac{1}{\bar{m}}\int_0^L \left(\rho ac/2\right)R^2\cos\theta\left[\left(d_1^2 + \frac{d_2^2}{2}\right) + \lambda_0\left(2d_1d_3 + d_2d_4\right)\right]\gamma \mathrm{d}x$$

$$f = \frac{1}{\bar{m}}\int_0^L \rho acR^2\cos\theta\left[d_1d_2 + \lambda_0\left(d_1d_4 + d_2d_3\right) + \lambda_0^2 d_3d_4\right]\gamma \mathrm{d}x$$
$$-\frac{1}{\bar{m}}\int_0^L \frac{\rho ac}{2}\left[\left(1-\beta_p^2\right)x + e_x\right]R\sin\theta\left(d_2 + \lambda_0 d_4\right)\gamma \mathrm{d}x \tag{A.6}$$

References

[1] Chopra I and Dugundji J 1979 *J. Sound Vib.* **63** 265–86
[2] Larsen J W and Nielsen S R K 2006 *Int. J. Nonlin. Mech.* **41** 629–43
[3] Larsen J W and Nielsen S R K 2007 *J. Sound Vib.* **299** 64–82
[4] Larsen J W, Iwankiewicz R and Nielsen S R K 2007 *Probabilist. Eng. Mech.* **22** 181–93
[5] Ramakrishnan V and Feeny B F 2011 In-plane nonlinear dynamics of wind turbine blades *Int. Design Engineering Technical Conf. & Computers and Information in Engineering Conf. (Washington, D C, USA, 28–31 August, 2011)* DETC 2011-4
[6] Kammer D C and Schlack A L Jr 1987 *J. Appl. Mech.-T. ASME* **54** 305–10
[7] Kammer D C and Schlack A L Jr 1987 *J. Vib. Acoust. Stress Reliab. Des.-T. ASME* **109** 138–43
[8] Young T H 1991 *J. Sound Vib.* **150** 435–46
[9] Asrar W 1991 *J. Sound Vib.* **150** 447–56
[10] Young T H and Liou G T 1993 *J. Sound Vib.* **164** 157–71
[11] Cai Y and Chen S S 1994 *J. Eng. Mech.* **120** 1786–02
[12] Tagata G 1995 *J. Sound Vib.* **185** 51–78
[13] Yang S M and Tsao S M 1997 *Comput. Struct.* **62** 643–51
[14] Chakraborty G and Mallik A K 1998 *Nonlinear Dynam.* **17** 301–24
[15] Öz H R and Pakdemırlı M 1999 *J. Sound Vib.* **227** 239–57
[16] Son I S, Uchiyama Y, Lacarbonara W and Yabuno H 2008 *Nonlinear Dynam.* **53** 129–38
[17] Wang X, Chen F Q and Zhou L Q 2009 *Nonlinear Dynam.* **56** 101–19
[18] Abolghasemi M and Jalali M A 2003 *Int. J. Nonlin. Mech.* **38** 739–51
[19] Younesian D and Esmailzadeh E 2010 *Nonlinear Dyn.* **60** 193–05
[20] Li B L, Song X G, He D X and An Y H 1999 *Structure Dynamics of Wind Turbines* (Beijing: Buaa Press)
[21] Li Y H, Li L, Liu Q K and Lv H W 2013 Dynamic characteristics of lag vibration of a wind turbine blade *Acta. Mech. Solida. Sin.* (accepted)

Journal of Physics: Conference Series **448** (2013) 012014 doi:10.1088/1742-6596/448/1/012014

Natural frequency analysis of fluid-conveying pipes in the ADINA system

L Wang, J Gan and Q Ni
Department of Mechanics, Huazhong University of Science and Technology, Wuhan 430074, China
Hubei Key Laboratory for Engineering Structural Analysis and Safety Assessment, Wuhan 430074, China

E-mail: ganjun23@126.com

Abstract. In this paper we present analysis and numerical simulations for the natural frequency of pipes conveying fluid in the ADINA system. The slender pipe structures are modelled as shells and the fluid flows are supposed to be three-dimensional (3D) and incompressible. The fluid and structural models are mechanically coupled on their interface through slip and displacement conditions. Several numerical examples for pipes shaped as different configurations show that the natural frequencies obtained based on ADINA codes are generally lower than those predicted by analytical or dynamic stiffness methods. In particular, for curved pipes conveying fluid with relatively high flow velocity, it is found that the evolution trend of natural frequencies with increasing flow velocity is similar as that predicted by the inextensible theory for curved pipes conveying fluid.

1. Introduction

The dynamical system of slender pipes conveying fluid is capable of displaying rich dynamical behaviour. As it is well known that, van der Pol's oscillator and the Lorenz equations are examples of systems the study of which has been instrumental in the development of modern dynamical theory [1, 2]. Now the model problem of pipes conveying fluid has established itself as another new paradigm in dynamics. This is a physically simple system, easily modelled by simple equation, and is a fairly easily realizable system with the possibility of theoretical and experimental investigation in parallel. Moreover, the model problem of pipes conveying fluid may be considered as the simplest fluid-structure interaction system [2]. The knowledge gained from the canonical problem of fluid-conveying pipes may radiate into other dynamic problems across applied mechanics [3]. Indeed, the dynamics of a pipe conveying fluid has been used in the concept of novel methods of marine propulsion, in the study of pulmonary collapse, and as medium for explaining the nature of turbulence and the dynamics of solar wind [2]. It is not surprising, therefore, that the literature on this topic is extensive and constantly expanding.

Certainly, some theoretical aspects of the problem have been studied extensively in the past decades. For instance, the vibration characteristics of straight pipes conveying fluid have been analyzed by many researchers. It would seem an easy task to extend the analysis of straight pipes conveying fluid to that of curved pipes. This is not so, however. In fact, in contrast to the systems discussed so far, there remained considerable uncertainty as to the vibration characteristics of curved pipes conveying fluid. The curved pipes shaped as circular arcs may be studied analytically. More complex L- and S-shaped configurations (or arbitrary ones), however, should be analysed by numerical methods such as finite-element method (FEM) [2].

In this paper, based on the FEM, numerical simulations are performed for the vibration characteristics of fluid-conveying pipes in the ADINA system. The ADINA system is a general-purpose finite element and finite volume code for the analysis of structures, fluid flows and fluid-structure interaction [4]. The objective of the numerical simulations is to valuate the capabilities of the ADINA code for analyzing the vibration characteristics of fluid-conveying pipes shaped with various configurations.

2. Modelling of fluid-conveying pipes in the ADINA system

2.1. Geometric model

In this paper, several typical configurations of fluid-conveying pipes with both ends simply supported will be considered. The systems of pipes shaped as straight, slightly curved and circular arcs will be studied. As a great deal of work has been done on the dynamics of straight pipes conveying fluid over the past 60 years or so, numerical simulations based on ADINA codes will be performed first, to demonstrate the validity of the solution procedure. In general, pipes may be curved and twisted into complex spatial forms. In this paper, however, reflecting the state of the art, curved pipes which initially lie within a given plane will be considered. For comparison convenience, only the in-plane vibration will be analyzed in the numerical results.

2.2. Assumptions

In the interest of analysing the vibration of fluid-conveying pipes in the transverse direction, only slender pipes will be studied. Thus, even for pipes with relatively thin wall, the lowest natural frequency will be only related to the beam-type mode. The internal fluid flow is assumed to be laminar and incompressible. Moreover, the friction between the fluid and pipe wall will be neglected according to the discussion by Paidoussis [5]. Thus, slip conditions between the internal fluid flow and the pipe wall may be used in the ADINA simulations.

2.3. Mechanical coupling on interface

The fluid and pipe structure in the ADINA system can be coupled through their interface. At fluid-structure interfaces, besides the slip conditions between the fluid flow and the pipe wall mentioned above, the compatibility of displacements/velocities must be satisfied. Since the internal fluid is flowing along the pipe length, the displacement compatibility may be viewed as a kinematic condition that can be realized along the normal direction of the interface.

2.4. Solution procedure

In the ADINA system, the presence of powerful analysis features for both solid and fluid models is essential for the accurate analysis of couple problems [4]. However, the coupling scheme is not readily available in either fluid or solid models. Thus, the well-organized core FSI coupling scheme should be used to link the two models. In order to handle the moving interface between fluids and solids, an arbitrary Lagrangian-Eulerian (ALE) formulation can be applied to the fluid. In the solution procedure, the direct computing method is used to solve the fully coupled equations of solid and fluid.

3. Results and discussion

In this section, some numerical examples that illustrate the capabilities of the ADINA system for predicting the vibration characteristics of fluid-conveying pipes will be presented.

The problems of fluid-conveying pipes, of course, have been extensively studied in both theoretical and experimental investigations. However, our purpose is to analyze the natural frequencies of such FSI systems by numerical simulations in the ADINA system. Here, it is of interest to discuss the effect on the natural frequency of the internal fluid flow. In the ADINA-FSI core scheme, the natural frequencies of the coupled system could not be calculated directly. If, however, an additional harmonic force is applied at the middle of the pipe, the frequency response curve may be calculated for each fluid velocity. From the frequency response curve, the natural frequency can be readily recognized.

For calculating convenience, the pipe systems considered are assumed to be slender and the dimensional parameters are chosen to be: pipe outer diameter D=52mm, inner diameter d=48mm, Young's modulus E=210GPa, Possion ratio ν = 0.3, pipe mass density ρ_p=7800kg/m^3 and fluid mass density ρ_f =1000kg/m^3.

In the following simulations, the pipe structures are all modeled using nine-node shell elements while the fluid domain is modelled using 3D four-node elements. The initial fluid velocity at the inlet of the pipe is prescribed at 0m/s. The increase of this fluid velocity is set to 0.1m/s. A transient analysis will be performed with a time step length of 0.1s.

3.1. Straight pipe conveying fluid

Let us analyse the vibrations of a straight pipe conveying fluid with simply supported boundary conditions first. The schematic view of the straight pipe is shown in Figure 1. For this geometrical pipe system, it has been verified theoretically and experimentally that the natural frequency would decrease with increasing fluid velocity.

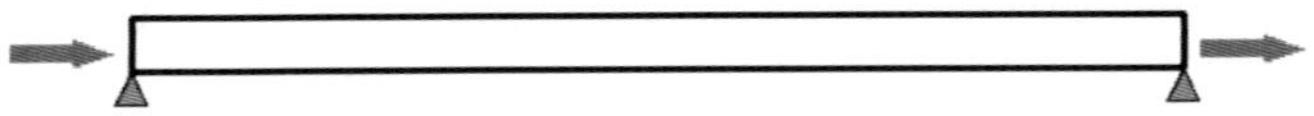

Figure 1. Schematic view of a straight pipe conveying fluid.

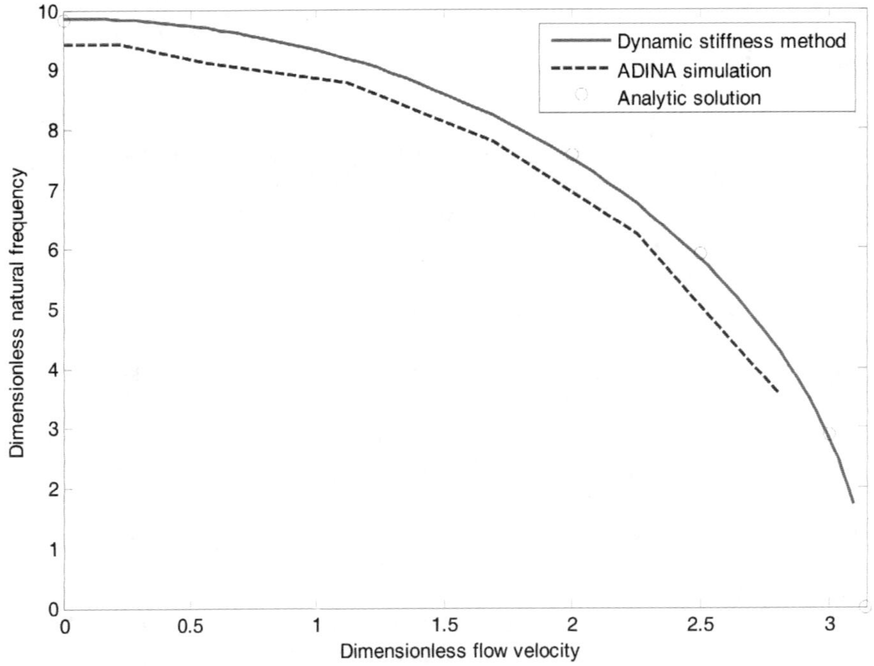

Figure 2. Natural frequencies as functions of the flow velocity for a pinned-pinned straight pipe.

In Figure 2, the calculations have been carried out for the first mode of a straight pipe with length L=1.2m. It can be seen that the natural frequencies decrease with increasing flow velocity. The results obtained using the ADINA code show reasonable agreement with those predicted by dynamic stiffness method (DSM) [6] and analytical method [5]. It is obvious that, moreover, the natural frequency predicted by ADINA system is lower than that obtained by dynamic stiffness method or analytical method. For sufficiently high flow velocity close to the critical value, the difference between the results predicted by the ADINA code and the other two methods becomes remarkable. In the case of sufficiently high flow velocity, it is suggested that maintaining numerical stability is not easy in the ADINA system.

3.2. Slightly curved pipe conveying fluid

Next, a slightly curved pipe shaped as circular arc will be analyzed. The slightly curved pipe model is shown in Figure 3. The length of the pipe is chosen to be L=1.2m. The curvature radius of the pipe is R=7.0m.

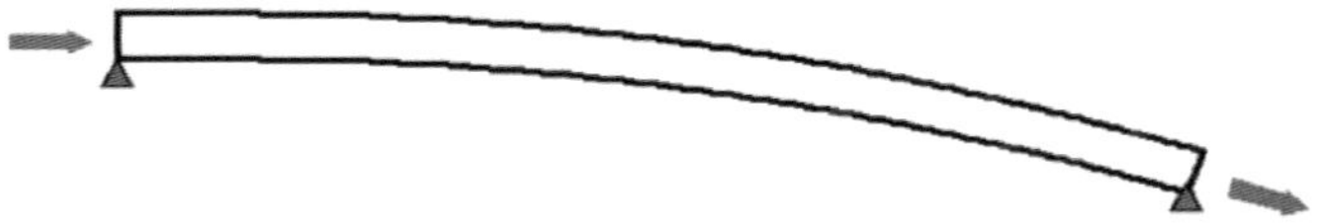

Figure 3. Schematic view of a slightly curved pipe conveying fluid.

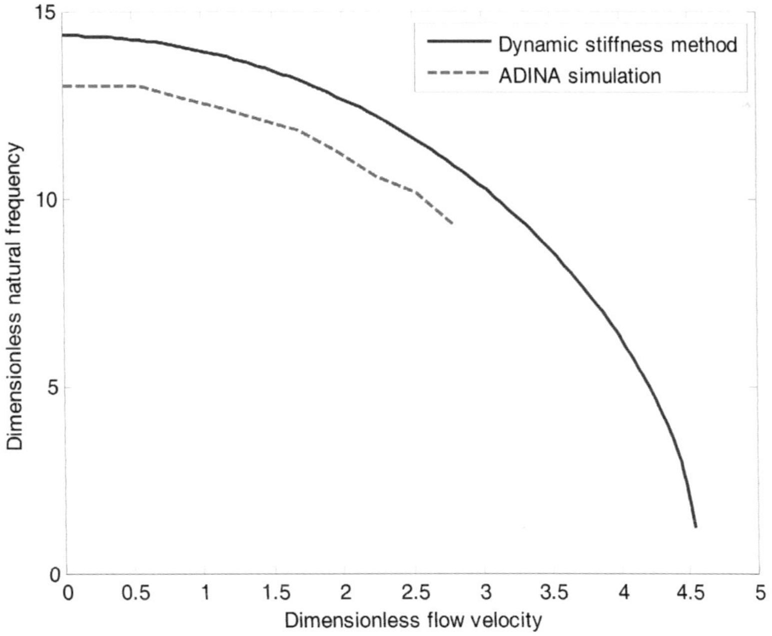

Figure 4. Natural frequencies as functions of the flow velocity for a slightly curved pipe with pinned-pinned boundary conditions.

In this case, the pipe is slightly curved; hence one expects the natural frequency would be higher than that of a straight pipe conveying fluid. The natural frequency of the system is shown in Figure 4. Here, comparing the ADINA results to the results predicted by dynamic stiffness method based on an "inextensible theory" [7, 8], it is noted the natural frequency predicted by ADINA system is, once again, lower than that predicted by the dynamic stiffness method. It should be pointed out that the inextensible theory for curved pipes conveying fluid presumes that the centreline of the pipe is essentially inextensible and all steady-state stress resultants are zero.

For higher flow velocity, similarly as before, numerical stability in the ADINA system could not maintain; and hence we have not obtained the corresponding natural frequencies.

3.3. Semi-circular curved pipe conveying fluid

It is then of interest to see how a semi-circular curved pipe behaves. For comparison purpose, the analytical method [7] and the dynamic stiffness method will be used to calculate the natural frequencies of the system. It should be mentioned that, for the dynamic stiffness method and analytical method, the equations of motion used are based on the inextensible theory initiated by Chen [7].

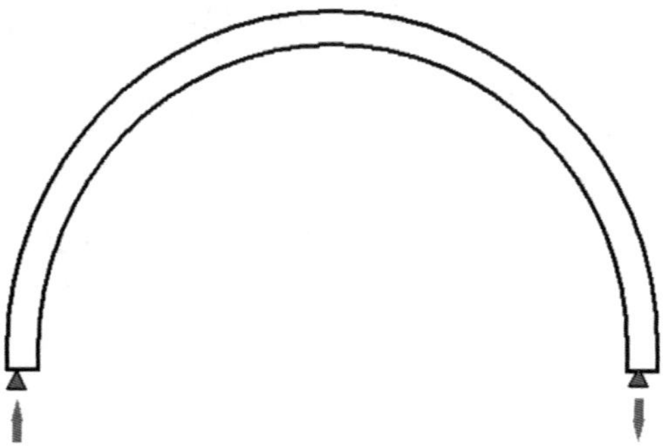

Figure 5. Schematic view of a semi-circular curved pipe conveying fluid.

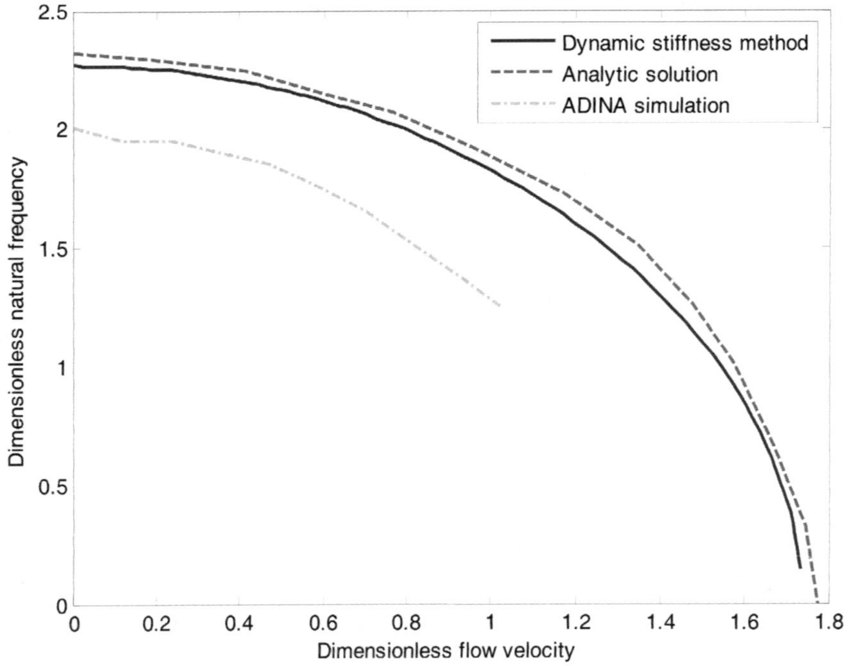

Figure 6. In-plane natural frequencies as functions of the flow velocity for a semi-circular curved pipe with pinned-pinned boundary conditions.

For a semi-circular curved pipe with both ends simply supported and curvature radius R=0.5m (see Figure 5), the evolution of the in-plane natural frequency with increasing dimensionless flow velocity is shown in Figure 6. As seen in this figure, the natural frequency is much lower than that predicted by the dynamic stiffness method or analytical method. For high flow velocity, similarly as before, the natural frequency could not be calculated by the ADINA code since the numerical stability is not easy to maintain.

3.4. Discussion

Clearly there are two significant characteristics from the ADINA system for predicting the vibrations of fluid-conveying pipes. The first is associated with the quantitative results of natural frequencies. The natural frequencies predicted by the ADINA system are much lower than those obtained by analytical method or dynamic stiffness method. For high flow velocity close to the critical value, the natural frequency could not be calculated correctly due to the numerical instability. The second is associated with the evolution trend of the natural frequencies with increasing flow velocity for curved pipes. In the results shown in Figures 4 and 6, the natural frequencies for curved pipes are shown to decrease with increasing flow velocity in the ADINA system. This trend agrees with that predicted by the analytical method and dynamic stiffness method. In these two figures, however, the analytical solutions or DSM solutions were all based on the inextensible theory. As discussed by Paidoussis [5] and Misra and Paidoussis [9], the inextensible theory and its conclusions regarding evolution trend of natural frequencies with increasing flow velocity may be not reliable.

A more reliable theory for curved pipe conveying fluid is the so-called “extensible theory” [9]. The extensible theory, as the name implies, do not make the assumption of that the centreline of the pipe is inextensible, and generally take into account the changes in form with increasing flow velocity, as well as the force generated thereby [5]. According to the results predicted by extensible theory [5, 9], the natural frequency of curved pipe may change slightly with increasing flow velocity. The curved pipe is stiffened by the flow-induced steady-state stresses (mainly centrifugally induced tension) at a rate. Thus, the flow-induced steady-state stresses may result in a stable system, no matter how high is the flow velocity.

4. Conclusions

This paper has initiated to model the vibration problem of fluid-conveying pipes in the ADINA system. By investigating the natural frequencies of pipes shaped as straight, slightly curved and semi-circular curved configurations, it is found that the natural frequencies of ADINA solutions are generally lower than those predicted by the analytical method and dynamic stiffness method. For curved pipes containing internal fluid flow, particularly the case of curved pipes conveying relatively high flow velocity, the evolution trend of natural frequencies with increasing fluid velocity is similar as that predicted by theoretical models based on inextensible theory.

Further studies require more comprehensive numerical simulations for a wider range of pipe configurations. Especially, fluid-conveying pipes shaped as various three-dimensional configurations need to be analyzed based on the ADINA code. Moreover, there exist some possible measures to improve numerical stability in the ADINA system. These and many more are some of the aspects of the problem should be considered in the near future.

Acknowledgments

The authors are grateful to the National Natural Science Foundation of China (11172107 and 11172109) and the Program for New Century Excellent Talents in University (NCET-11-0183) for financial support in this study.

References

[1] Guckenheimer J and Holmes P 1983 *Nonlinear oscillations, dynamical systems, and bifurcations of vector fields* (New York: Springer)

[2] Paidoussis M P and Li G X 1993 *J. Fluids Struct.* **7** 137
[3] Paidoussis M P 2008 *J. Sound Vib.* **310** 462
[4] Zhang H, Zhang X L, Ji S H, Guo Y H, Ledezma G, Elabbasi G and deCougny H 2003 *Comput. Struct.* **81** 1071
[5] Paidoussis M P 1998 *Fluid–Structure Interactions: Slender Structures and Axial Flow* Vol 1 (London: Academic Press Limited)
[6] Dai H L, Wang L, Qian Q and Gan J 2012 *Appl. Math. Comput.* **219** 2453
[7] Chen S S 1972 *Nucl. Eng. Des.* **23** 29
[8] Misra A K, Paidoussis M P and Van K S 1988 *J. Fluids Struct.* **2** 221
[9] Misra A K, Paidoussis M P and Van K S 1988 *J. Fluids Struct.* **2** 245

Journal of Physics: Conference Series **448** (2013) 012015 doi:10.1088/1742-6596/448/1/012015

Nonlinear dynamic characteristics of a flexible rotor system with local rub-impact

H. Ma[1], X. Y. Tai[1], H. L. Yi[2], S. Lv[1], B. C. Wen[1]
[1] School of Mechanical Engineering and Automation, Northeastern University, Shenyang 110819, China
[2] Shenyang Blower Works Group Corporation, Shenyang, 110869, China

E-mail: mahui_2007@163.com

Abstract. In this paper, nonlinear dynamic characteristics of a single span rotor system with two discs are investigated under fixed-point and local arc rub-impact conditions. A twenty-degree-of-freedom model considering the gyroscopic effect is developed, the simple Coulomb friction model and piecewise linear spring model to describe the contact between the rotor and the stator. The vibration chracteristics of the rotor system with two types of rub-impact forms are analyzed with respect to the effects of rotating speed by using spectrum cascades, vibration waveforms, orbits and frequency spectra. The simulation results show that different rotor motions appear, such as P1(period-1), P2, P3 and P4 with the increasing rotating speed under two rub-impact conditions. And some combination frequency components and continuous spectra appear under the local arc rub-impact condition, which are different from those under the fixed-point rub-impact condition.

1. Introduction

Due to the increasing demand for high speed and high efficiency, the clearance between the rotor and stator in rotating machinery is becoming smaller and smaller. As a result, the rub-impact, which refers to the contact between the rotating and non-rotating structures in a machine, has become one of the most common damaging malfunctions in rotating machinery. Many works on different rub-impact forms, such as point, local annular and full annular rub-impacts, have been extensively studied [1].

Many researches on the fixed-point rub-impact have been carried out by Ma et al., Han et al., Lahriri et al., Hu et al. and many others. Ma et al. [2,3] analyzed nonlinear dynamic characteristics of a flexible rotor system with fixed-point rubbing by numerical simulation and experimental verification. Han et al. [4] developed a FE (finite element) model of the rotor system and analyzed different rotor motions caused by rub-impact occurring at fixed limiters. Lahriri et al. [5,6] proposed a new unconventional backup bearing design (similar to four point rubbing device) in order to reduce the rub related severity in friction and center the rotor at impact events. The result shows that the rotor with impacts is forced to the center of the backup bearing and the lateral motion is mitigated. Hu et al. [7] analyzed the features of sharp rubbing between the rotor and stator in a rotor system by experiments. Their results show that in the early rubbing stage, there are the sub-harmonics vibration of the order 1/3 and 2/3 obviously.

[1] Corresponding author (H. Ma). Tel.: +86 24 83684491; fax: +86 24 83679731.

Many efforts have been devoted to study the local rubbing fault, such as Muszynska et al., Abuzaid et al., Chu et al., and Roques et al. Muszynska et al. [8] discussed chaotic response of unbalance rotor-bearing-stator systems with local rubbing. The results show that at low rotational speeds, the rotor's response exhibits 1*X* (*X* denotes rotating frequency) vibration as the lowest frequency component. At a speed roughly equal to twice the rotor's first lateral natural frequency, a 1*X*/2 sub-synchronous vibration component appears, and continuous until the rotational speed reaches 3 times the rotor's first natural frequency. Then, the 1*X*/2 is replaced by 1*X*/3 vibration as the lowest frequency component. Abuzaid et al. [9] investigated the effect of partial rotor-stator rubbing both experimentally and analytically. The results show that the features of light rubbing include 1*X*, 2*X* and 3*X*, whereas severe rubbing include 1*X*/3 and 2*X*/3. Chu et al. [10] studied the nonlinear vibration of a rub-impact rotor system and observed a variety of periodic and chaotic vibrations. The experiments show that the system motion generally contains multiple harmonic components, such as 2*X*, 3*X*, and 1/2-based fractional harmonic components, such as 1*X*/2 and 3*X*/2. Under some special conditions, the 1/3-based fractional harmonic components can also be observed. Roques et al. [11] presented a rotor-stator model of a turbogenerator and investigated rotor-stator rubbing caused by an accidental blade-off imbalance. Thereinto, the rotor system is modeled using FE method, rotor-stator rubbing is simulated using node-to-line contact and the highly nonlinear equations due to contact conditions are solved through an explicit prediction-correction time-marching procedure combined with the Lagrange multiplier approach.

Aiming at full annular rubbing, Dai et al. [12,13] adopted the simple Coulomb friction model and piecewise linear spring model to describe the contact between the rotor and the stationary stop, established the dynamic equations of the rotor that does precession roughly enough to rub with the displacement stop in the support and analyzed nonlinear vibration responses in the process of rubbing by numerical simulation. Choi [14] investigated the whirling motion of the rotor with full annular rub. His study shows that the effects of the friction coefficient and the eccentricity of the rotor can explain the onset of backward rolling and backward slipping as the rotor speed increases or decreases. Jiang et al. [15] analyzed the stability of the full annular rub solutions of a modified Jeffcott rotor with a given rotor-stator clearance and cross-coupling influences. Adopting a linear spring model to model the local impact-deformation phenomenon, Xu et al. [16] studied the synchronous full annular rub motion of a Jeffcott rotor and its dynamic stability behaviour in detail.

From experiments on aero-engine test rigs, it has been observed that rubbings often only occur at fixed points or local arc area along the radial direction if they are caused by the deformation of compressor cases [17], which is called fixed-point or local arc rub-impact in this paper. In our study, a twenty-degree-of-freedom model of a flexible rotor system considering the gyroscopic effect is established. Vibration responses are simulated under fixed-point and local arc rub-impact conditions. The results will give deep insight into the local rub-impact mechanism.

2. Fixed-point and local arc rub-impact models

Fixed-pint rub-impact is that the rotor contacts a fixed point of the stator once in one cycle, which usually occurs when the stator stiffness is greater than that of the rotor system or bumps caused by the stator deformation appear. Schematic diagram of the fixed-point rub-impact is shown in figure 1(a). In the figure, ω is rotating speed, F_{rx} and F_{ry} are normal and tangential rub-impact forces respectively and their expressions are as follows

$$\begin{cases} F_{rx} = -k_r(x-c)H(x-c) \\ F_{ry} = f_r F_{rx} = -f_r k_r (x-c)H(x-c) \end{cases}, \tag{1}$$

where c is an initial clearance between the rotor and the stator, k_r is the stator stiffness in the rub-impact direction, f_r is friction coefficient at rub-impact position, H is Heaviside function and its expression can be written as

$$H(x-c)=\begin{cases}0 & \text{if } x<c \\ 1 & \text{if } x\geq c\end{cases}.$$

Schematic diagram of the local arc rub-impact is shown in figure 1(b). In the figure, o is the center of the stator, o' is the center of the rotor, β is the angle between oo' and the positive x-axis, α_0 and α_1 are start and end angles between the stator and the positive x-axis, F_n and F_t are normal and tangential rub-impact forces respectively and their expressions are as follows

$$\begin{cases}F_n=(r-c)k_r \\ F_t=f_rF_n\end{cases} \quad (r\geq c \text{ and } \alpha_0\leq\beta\leq\alpha_1), \tag{2}$$

where $r=\sqrt{x^2+y^2}$ is the radial displacement of the rotor. The component forces F_{rx} and F_{ry} of rub-impact force in x and y directions can be expressed as follows

$$\begin{Bmatrix}F_{rx} \\ F_{ry}\end{Bmatrix}=-H(r-c)\frac{(r-c)k_r}{r}\begin{pmatrix}1 & -f_r \\ f_r & 1\end{pmatrix}\begin{Bmatrix}x \\ y\end{Bmatrix} \quad (\alpha_0\leq\beta\leq\alpha_1), \tag{3}$$

where $H(r-c)=\begin{cases}0 & \text{if } r<c \\ 1 & \text{if } r\geq c\end{cases}$.

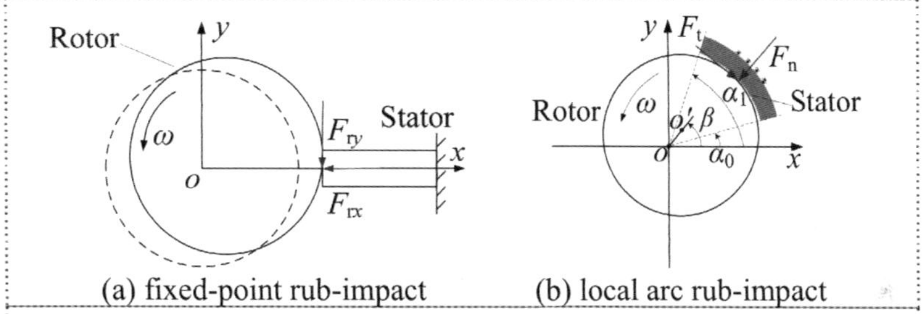

Figure 1. Schematic diagrams of fixed-point and local arc rub-impacts

3. The motion differential equations of the rotor system with rub-impact

In order to study the rub-impact fault efficiently, a mathematical model of the rotor-bearing system, which is depicted in Fig. 1, is simplified according to the following assumptions.

(a) The movements of the rotor in torsional and axial directions are negligible; the journals, coupling and discs are simulated by five lumped mass points and the corresponding points are connected by massless shaft sections of lateral stiffness; each point has four degrees of freedom including two rotation and two translation freedom, in which m_i (i=1,2,3,4,5) are lumped masses.

(b) The bearings are modelled ideally by linear stiffness and damping, where k_{blx}, k_{bly}, c_{blx}, c_{bly} and k_{brx}, k_{bry}, c_{brx}, c_{bry} are stiffnesses and dampings of the left and right bearings in x and y directions, respectively.

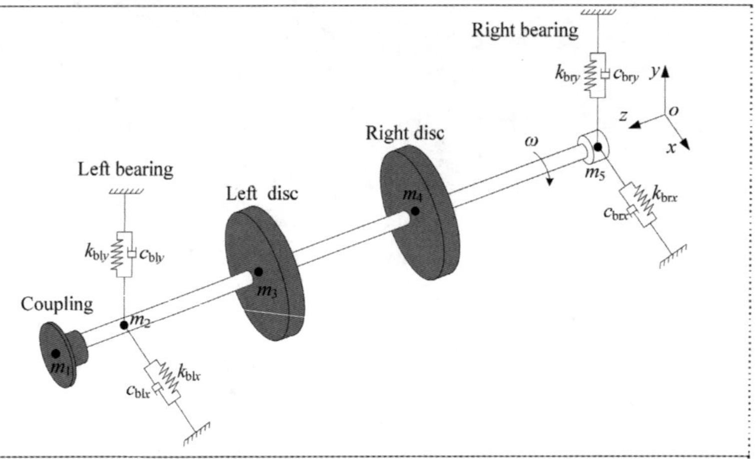

Figure 2. Schematic diagram of a rotor system

Assuming that rub-impact occurs at the left disc and the motion differential equations of the rotor-bearing system with twenty degrees of freedom can be deduced as

$$\boldsymbol{M}\ddot{\boldsymbol{q}} + (\boldsymbol{G}+\boldsymbol{C})\dot{\boldsymbol{q}} + \boldsymbol{K}\boldsymbol{q} = \boldsymbol{F}_{\mathrm{u}} + \boldsymbol{F}_{\mathrm{r}}, \tag{4}$$

$$\boldsymbol{M} = \begin{bmatrix} \boldsymbol{M}_x & 0 \\ 0 & \boldsymbol{M}_y \end{bmatrix}, \ \boldsymbol{M}_x = \boldsymbol{M}_y = \mathrm{diag}[m_1, J_{\mathrm{d1}}, m_2, J_{\mathrm{d2}}, m_3, J_{\mathrm{d3}}, m_4, J_{\mathrm{d4}}, m_5, J_{\mathrm{d5}}], \tag{5}$$

$$\boldsymbol{G} = \omega\boldsymbol{J} = \omega\begin{bmatrix} 0 & \boldsymbol{J}_1 \\ -\boldsymbol{J}_1^{\mathrm{T}} & 0 \end{bmatrix}, \ \boldsymbol{J}_1 = \mathrm{diag}[0, J_{\mathrm{p1}}, 0, J_{\mathrm{p2}}, 0, J_{\mathrm{p3}}, 0, J_{\mathrm{p4}}, 0, J_{\mathrm{p5}}], \tag{6}$$

where $J_{\mathrm{p}i}$ and $J_{\mathrm{d}i}$ (i=1,2…5) are the polar moment of inertia and the diametral moment of inertia of lumped points, respectively.

$$\boldsymbol{K} = \begin{bmatrix} \boldsymbol{K}_x & 0 \\ 0 & \boldsymbol{K}_y \end{bmatrix}, \tag{7}$$

$$\boldsymbol{K}_x = \begin{bmatrix}
k_{11} & k_{12} & k_{13} & k_{14} & 0 & 0 & 0 & 0 & 0 & 0 \\
k_{12} & k_{22} & k_{23} & k_{24} & 0 & 0 & 0 & 0 & 0 & 0 \\
k_{13} & k_{23} & k_{33}+k_{\mathrm{blx}} & k_{34} & k_{35} & k_{36} & 0 & 0 & 0 & 0 \\
k_{14} & k_{24} & k_{34} & k_{44} & k_{45} & k_{46} & 0 & 0 & 0 & 0 \\
0 & 0 & k_{35} & k_{45} & k_{55} & k_{56} & k_{57} & k_{58} & 0 & 0 \\
0 & 0 & k_{36} & k_{46} & k_{56} & k_{66} & k_{67} & k_{68} & 0 & 0 \\
0 & 0 & 0 & 0 & k_{57} & k_{67} & k_{77} & k_{78} & k_{79} & k_{7,10} \\
0 & 0 & 0 & 0 & k_{58} & k_{68} & k_{78} & k_{88} & k_{89} & k_{8,10} \\
0 & 0 & 0 & 0 & 0 & 0 & k_{79} & k_{89} & k_{99}+k_{\mathrm{brx}} & k_{9,10} \\
0 & 0 & 0 & 0 & 0 & 0 & k_{7,10} & k_{8,10} & k_{9,10} & k_{10,10}
\end{bmatrix}, \tag{8}$$

$$\boldsymbol{K}_y = \begin{bmatrix} k_{11} & -k_{12} & k_{13} & -k_{14} & 0 & 0 & 0 & 0 & 0 & 0 \\ -k_{12} & k_{22} & -k_{23} & k_{24} & 0 & 0 & 0 & 0 & 0 & 0 \\ k_{13} & -k_{23} & k_{33}+k_{bly} & -k_{34} & k_{35} & -k_{36} & 0 & 0 & 0 & 0 \\ -k_{14} & k_{24} & -k_{34} & k_{44} & -k_{45} & k_{46} & 0 & 0 & 0 & 0 \\ 0 & 0 & k_{35} & -k_{45} & k_{55} & -k_{56} & k_{57} & -k_{58} & 0 & 0 \\ 0 & 0 & -k_{36} & k_{46} & -k_{56} & k_{66} & -k_{67} & k_{68} & 0 & 0 \\ 0 & 0 & 0 & 0 & k_{57} & -k_{67} & k_{77} & -k_{78} & k_{79} & -k_{7,10} \\ 0 & 0 & 0 & 0 & -k_{58} & k_{68} & -k_{78} & k_{88} & -k_{89} & k_{8,10} \\ 0 & 0 & 0 & 0 & 0 & 0 & k_{79} & -k_{89} & k_{99}+k_{bry} & -k_{9,10} \\ 0 & 0 & 0 & 0 & 0 & 0 & -k_{7,10} & k_{8,10} & -k_{9,10} & k_{10,10} \end{bmatrix}. \quad (9)$$

The matrix elements of $\boldsymbol{K}_x$ and $\boldsymbol{K}_y$ are calculated as follows:

$$\begin{cases} k_{11}=a_{11} \\ k_{12}=a_{21} \\ k_{13}=-a_{11} \\ k_{14}=a_{21} \end{cases}, \begin{cases} k_{22}=l_1a_{21}-a_{31} \\ k_{23}=-a_{21} \\ k_{24}=a_{31} \end{cases}, \begin{cases} k_{33}=a_{11}+a_{12} \\ k_{34}=-a_{21}+a_{22} \\ k_{35}=-a_{12} \\ k_{36}=a_{22} \end{cases}, \begin{cases} k_{44}=l_1a_{21}-a_{31}+l_2a_{22}-a_{32} \\ k_{45}=-a_{22} \\ k_{46}=a_{32} \end{cases}, \begin{cases} k_{55}=a_{12}+a_{13} \\ k_{56}=-a_{22}+a_{23} \\ k_{57}=-a_{13} \\ k_{58}=a_{23} \end{cases},$$

$$\begin{cases} k_{66}=l_2a_{22}-a_{32}+l_3a_{23}-a_{33} \\ k_{67}=-a_{23} \\ k_{68}=a_{33} \end{cases}, \begin{cases} k_{77}=a_{13}+a_{14} \\ k_{78}=-a_{23}+a_{24} \\ k_{79}=-a_{14} \\ k_{7,10}=a_{24} \end{cases}, \begin{cases} k_{88}=l_3a_{23}-a_{33}+l_4a_{24}-a_{34} \\ k_{89}=-a_{24} \\ k_{8,10}=a_{34} \end{cases}, \begin{cases} k_{99}=a_{14} \\ k_{9,10}=-a_{24} \\ k_{10,10}=l_4a_{24}-a_{34} \end{cases},$$

and

$$\begin{cases} a_{1i}=\dfrac{12EI}{l_i^3} \\ a_{2i}=\dfrac{1}{2}l_ia_{1i}, i=1,2,3,4, \\ a_{3i}=\dfrac{1}{6}l_i^2a_{1i} \end{cases} \quad (10)$$

in which E, l_i and I are the Young's modulus of elasticity, the distance between every two consecutive lumped mass points and the area moment of inertia respectively.

$$\boldsymbol{C}=\boldsymbol{C}_1+\boldsymbol{C}_2, \quad (11)$$

$$\boldsymbol{C}_1=\alpha\boldsymbol{M}+\beta\boldsymbol{K}, \quad (12)$$

$$\alpha=\frac{60(\omega_{n2}\xi_1-\omega_{n1}\xi_2)\omega_{n1}\omega_{n2}}{\pi(\omega_{n2}^2-\omega_{n1}^2)}, \beta=\frac{\pi(\omega_{n2}\xi_2-\omega_{n1}\xi_1)}{15(\omega_{n2}^2-\omega_{n1}^2)}, \quad (13)$$

where ω_{n1} and ω_{n2} are the first and second order natural frequencies (r/min); ξ_1 and ξ_2 refer to the first and second order modal damping ratios, respectively.

$$\boldsymbol{C}_2=\mathrm{diag}[0,0,c_{blx},0,0,0,0,0,0,0,0,0,c_{bly},0,0,0,0,0,0,0] \quad (14)$$

$$\boldsymbol{q}=[x_1,\theta_{y1},x_2,\theta_{y2},x_3,\theta_{y3},x_4,\theta_{y4},x_5,\theta_{y5},y_1,\theta_{x1},y_2,\theta_{x2},y_3,\theta_{x3},y_4,\theta_{x4},y_5,\theta_{x5}]^{\mathrm{T}}, \quad (15)$$

where x_i, y_i, θ_{xi} and θ_{yi} (i=1,2,3,4,5) are the displacements in x and y directions and angles of orientation associated with the x and y axes, respectively.

$$\boldsymbol{F}_{\mathrm{u}} = [0,0,0,0,m_3 e\omega^2 \cos(\omega t + \varphi_1),0,m_4 e\omega^2 \cos(\omega t + \varphi_2),0,0,0,\\ 0,0,0,0,m_3 e\omega^2 \sin(\omega t + \varphi_1),0,m_4 e\omega^2 \sin(\omega t + \varphi_2),0,0,0]^{\mathrm{T}}, \tag{16}$$

where m_3e, m_4e, φ_1 and φ_2 denote the eccentricity of unbalance mass of the left and right discs, the initial phase angles of eccentricity in left and right discs, respectively.

$$\boldsymbol{F}_{\mathrm{r}} = [0,0,0,0,F_{\mathrm{r}x},0,0,0,0,0,0,0,0,0,F_{\mathrm{r}y},0,0,0,0,0]^{\mathrm{T}} \tag{17}$$

in which $F_{\mathrm{r}x}$ and $F_{\mathrm{r}y}$ are rub-impact forces of the left disc in x and y directions.

The model parameters of the rotor and the bearing are listed in Table 1. The first and second natural frequencies can be determined to be about 27.27 Hz and 98.48 Hz. Eq. (4) is solved by using numerical methods and Newmark-β integration method is adopted because it is a kind of robust algorithm for solving nonlinear equations in the time domain. In this paper, the spectrum cascade is used to exhibit continuous changes of the frequency components of the rotor-bearing system, rotor orbit is used to show the axis trace moving direction and vibration waveform to indicate time domain features under some rotating speeds.

Table 1. Model parameters of the rotor-bearing system

m_1,m_2,m_3,m_4,m_5 (kg)	0.0439, 0, 0.741, 0.741, 0
J_{p1},J_{d1} (kg·m^2)	2.42×10^{-6}, 2.56×10^{-6}
J_{p2},J_{d2}(kg·m^2)	0, 0
J_{p3},J_{d3}(kg·m^2)	4.09×10^{-4}, 2.4×10^{-4}
J_{p4},J_{d4}(kg·m^2)	4.09×10^{-4}, 2.4×10^{-4}
J_{p5},J_{d5}(kg·m^2)	0, 0
$k_{\mathrm{bl}x},k_{\mathrm{bl}y},k_{\mathrm{br}x},k_{\mathrm{br}y}$ (N/m)	1×10^5,2×10^5,2×10^8,5×10^8
$c_{\mathrm{bl}x},c_{\mathrm{bl}y},c_{\mathrm{br}x},c_{\mathrm{br}y}$ (N·s/m)	100,200,1000,2000
ξ_1 , ξ_2	0.02, 0.04
m_3e, m_4e (kg·m)	1.56×10^{-4}
φ_1, φ_2(rad)	0, 0

4. Numerical simulation of fixed-point and local arc rub-impacts

4.1. Fixed-point rub-impact simulation

Material parameters of rotor are as follows: Young's modulus of elasticity E=207 GPa, Poisson ratio υ=0.3. The stator stiffness k_{r}=8 MN/m and the clearance between the rotor and the stator c=0.1 mm, friction coefficient f_{r}=0.3. Spectrum cascade of the system with rotating speed is shown in figure 3. In the figure $1X$ denotes rotating frequency.

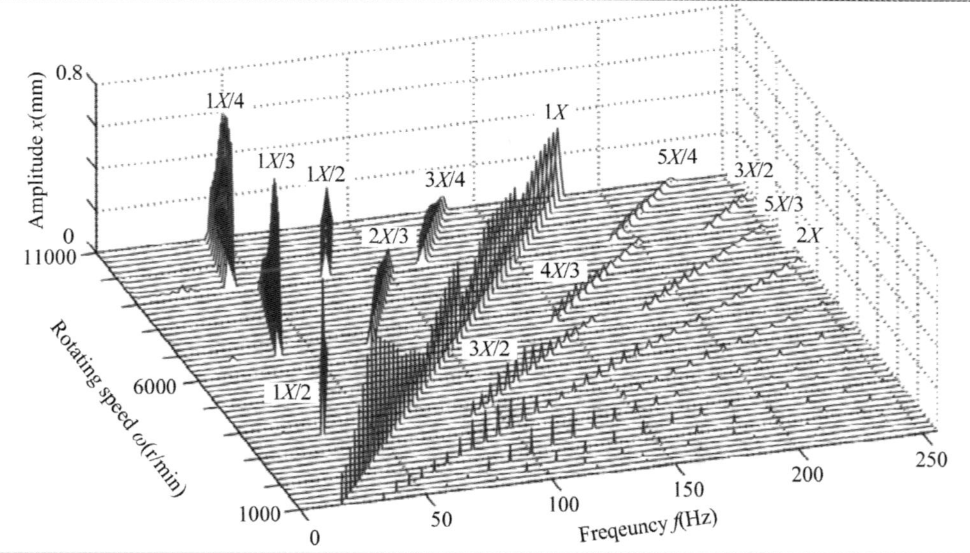

Figure 3. Spectrum cascade of the left disc with fixed-point rub-impact

From figure 3 it can be seen that the frequency feature includes 1*X*, 2*X* in the range of $\omega\in$[1000, 3500] r/min, which show system motion is P1 (period-1); 1/2 fractional harmonic components, such as 1*X*/2, 3*X*/2, 5*X*/2, etc., in the range of $\omega\in$[3500, 6300] r/min; 1/3 fractional harmonic components in the range of $\omega\in$[6300, 9200] r/min; 1/4 fractional harmonic components in the range of $\omega\in$[9200, 10000] r/min. The detailed vibration responses at ω=1000, 5000, 8500 r/min are shown in Fig. 4, which shows, from the left to the right, vibration waveforms, orbits and spectra. In the orbits, the black line denotes the orbits without rub-impact and red line the orbits with rub-impact.

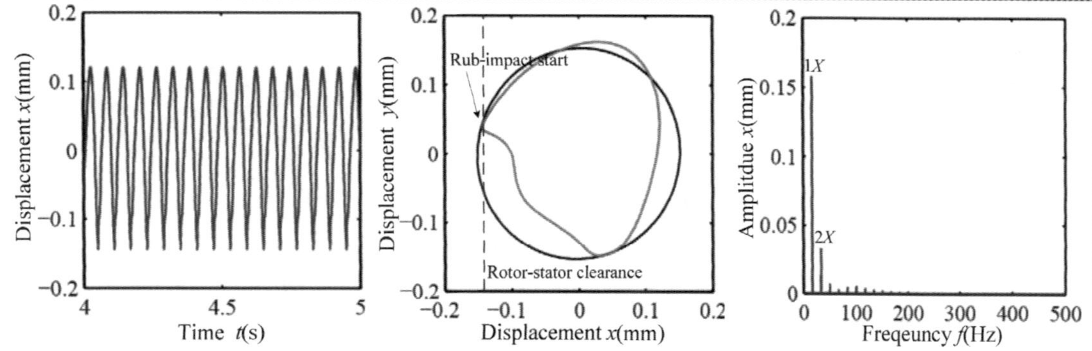

(a) ω=1000 r/min

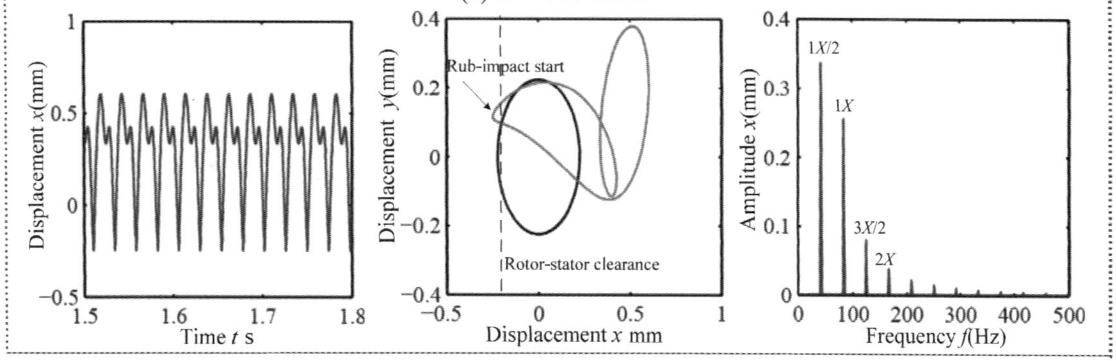

(b) ω=5000 r/min

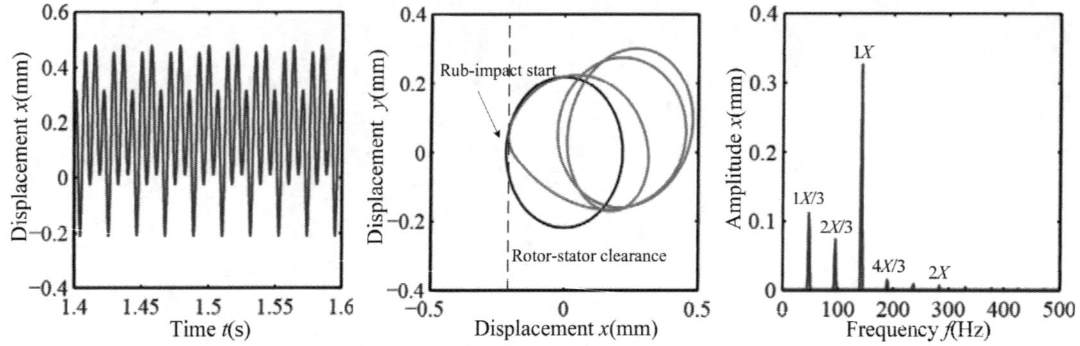

(c) ω=8500 r/min

Figure 4. Vibration response of the system at ω=1000,5000, 8500 r/min

4.2. Local arc rub-impact simulation

The clearance between the rotor and the stator c=0.14 mm and rub-impact regional angle $\alpha=\alpha_1$-α_0= 30° and other parameters are the same as those in section 4.1. Spectrum cascade of the system with rotating speed is shown in figure 5. As a whole, the system motions (P1, P2, P3 and P4) with local arc rub-impact are similar to those under fixed-point rub-impact condition. However frequency features of the former, especially the transition regions between different motions, are more complicated than those of the latter. Some continuous spectra appear in the ranges of $\omega \in$[1800,2000], [3000,3800], [5000,6000] and [8800,9200] r/min, which show the complexity of the local arc rub-impact.

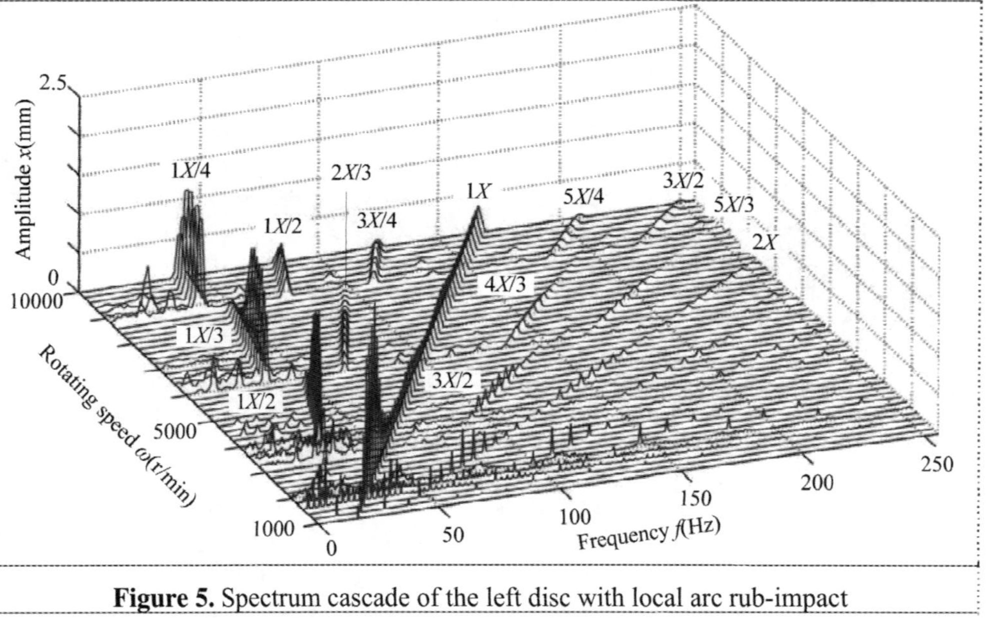

Figure 5. Spectrum cascade of the left disc with local arc rub-impact

The detailed vibration responses at ω=2000, 4000, 6000 r/min are shown in Fig. 6. Combination frequency components appear and the system motion is quasi-periodic at ω=2000 r/min. 1/2 fractional harmonic components, such as 1X/2, 3X/2, appear at ω=4000 r/min, which show system motion is P2. Low frequency components with big amplitudes and continuous spectra appear at ω=6000 r/min.

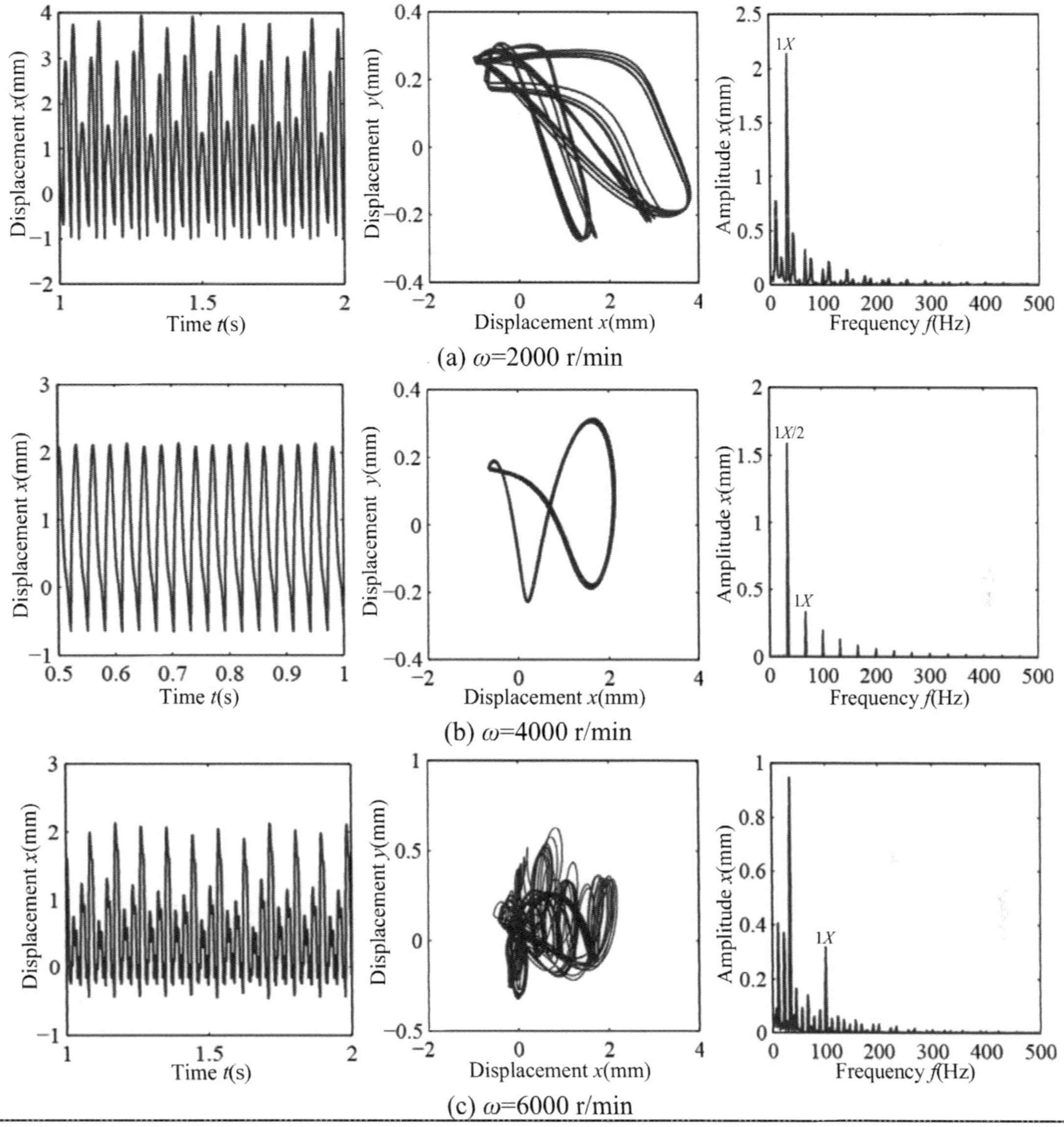

Figure 6. Vibration response of the system at ω=2000,4000, 6000 r/min

5. Conclusions

In this paper, a twenty-degree-of-freedom model of a flexible rotor system considering the gyroscopic effect is developed. The system vibration responses are analyzed under the fixed-point and local arc rub-impact conditions by numerical simulation. Some conclusions drawn from the study can be summarized as follows.

Under the fixed-point rub-impact condition, system motion is from period-1 through period-2 and period-3 to period-4 with the increase of the rotating speed. On the whole, the motion forms under the local rub-impact condition are similar to those under the fixed-point rub-impact, however, some combination frequency components and continuous spectra appear at the transition regions between different motions. These features show that the local rub-impact is more complicated than fixed-point rub-impact.

Acknowledgments

We are grateful to the China Natural Science Funds (NSFC, Grant No. 50805019), the Fundamental Research Funds for the Central Universities (Grant No. N100403008) and the Program for New Century Excellent Talents in University (Grant No. NCET-11-0078) for providing financial support for this work.

References

[1] Wen B C, Wu X H, Ding Q and Han Q K 2004 *Theory and experiment of nonlinear dynamics for rotating machinery with faults* (Beijing: Science Press)

[2] Ma H, Tai X Y, Sun J and Wen B C 2001 Analysis of dynamic characteristics for a dual-disk rotor system with single rub-impact *Advanced Science Letters* **4** 2782-89

[3] Ma H, Shi C Y, Han Q K and Wen B C 2012 Fixed-point rubbing fault characteristic analysis of a rotor system based on contact theory *Mech Syst Signal Pr Preprint* j.ymssp.2012.10.009

[4] Han Q K, Zhang Z W, Wen B C 2008 Periodic motions of a dual-disc rotor system with rub-impact at fixed limiter *P I Mech Eng C-J Mec* **222** 1935-46.

[5] Lahriri S, Weber H I, Santos I F, et al. 2012 Rotor-stator contact dynamics using a non-ideal drive-Theoretical and experimental aspects *J Sound Vib* **331** 4518-36.

[6] Lahriri S, Santos I F, Weber H I, et al. 2012 On the nonlinear dynamics of two types of backup bearings-theoretical and experimental aspects. *J Eng Gas Turb Power* **134** 112503-1-13.

[7] Hu N Q, Zhang Y, Liu Y Z, et al. 2002 Experimental research on vibration characteristics of sharp rub-impact between rotor and stator of a rotor system *China Mechanical Engineering* **13** 777-9

[8] Muszynaka A, Goldman P 1995 Chaotic responses of unbalanced rotor/bearing/stator system with looseness or rubs *Chaos Soliton Fract* **5** 1683-1704

[9] Abuzaid M A, Eleshaky M E, Zedan M G 2009 Effect of partial rotor-to-stator rub on shaft vibration *J Mech Sci Technol* **23** 170-182

[10] Chu F L, Lu W X 2005 Experimental observation of nonlinear vibrations in a rub–impact rotor system *J Sound Vib* **283** 621-643

[11] Roques S, Legrand M, Cartraud P, et al. 2010 Modeling of a rotor speed transient response with radial rubbing *J Sound Vib* **329** 527-546

[12] Dai X, Zhang X, Jin X 2001 The partial and full rubbing of a flywheel rotor-bearing-stop system *International Journal of Mechanical Sciences* **43** 505-519

[13] Dai X, Jin Z, Zhang X 2002 Dynamic behavior of the full rotor/stop rubbing: numerical simulation and experimental verification *J Sound Vib* **251** 807-822

[14] Choi Y S 2002 Investigation on the whirling motion of full annular rotor rub *J Sound Vib* **258** 191-198.

[15] Jiang J, Ulbrich H 2001 Stability analysis of sliding whirl in a nonlinear Jeffcott rotor with cross coupling stiffness coefficients *Nonlinear Dynamics* **24** 269-283

[16] Xu b, Xu W N, Zhang W 2006 Study of synchronous full annular rub of Jeffcott rotor and its dynamic stability *Journal of Fudan University* (*Natural Science*) **45** 148-154.

[17] Wang D Y 1995 Research and extract of vibration features on rubbing between rotor and stator for aeroengine PhD Thesis (Beijing: University of Aeronautics and Astronautics)

AUTHOR INDEX

Institute of Physics
Dirac House, Temple Back
Bristol BS1 6BE UK

CURRAN ASSOCIATES INC
proceedings.com